LOVE, ANGER & BETRAYAL

LOVE, ANGER & BETRAYAL

JUST STOP OIL'S YOUNG CLIMATE CAMPAIGNERS

JONATHON PORRITT

ANTHONY EYRE
MOUNT HOUSE PRESS

First published in the UK in 2025 by Anthony Eyre
Mount House Press, 23 High Street, Cricklade, Wiltshire SN6 6AP
www.anthonyeyre.com

Text © 2025 Jonathon Porritt

ISBN 9781912945542

All rights reserved. No part of this publication may be reproduced in any form or by any means – electronic, mechanical, photocopying, recording, or otherwise – or stored in any retrieval system of any nature without prior written permission from the copyright holders. Jonathon Porritt has asserted his moral right to be identified as the author of this work in accordance with the Copyright, Designs and Patents Act of 1988.

A CIP catalogue record for this book is available from the British Library.

Every reasonable effort has been made to trace copyright-holders of material reproduced in this book, but if any have been inadvertently overlooked the publishers would be glad to hear from them.

Design and typesetting by Anthony Eyre

Printed in the UK by Short Run Press, Exeter

10 9 8 7 6 5 4 3 2 1

JUST STOP OIL STATEMENT ON *LOVE, ANGER & BETRAYAL*

Love, Anger & Betrayal is written by Jonathon Porritt, working together with 26 young climate activists involved in Just Stop Oil. The opinions expressed in it are those of the author and of the interviewees (in their stand-alone extracts), not those of Just Stop Oil.

Just Stop Oil has been happy to support Jonathon Porritt in this endeavour, but he has played no part in the internal deliberations/decision-making of Just Stop Oil, and writes as a completely independent commentator on a wide range of sustainability issues today.

Jonathon Porritt has financed the project himself, and will be launching a crowd-funder to help cover production, printing and publicity costs. Once those costs have been covered, proceeds from sales will go to support the work of direct action climate campaigning.

Dedicated to my Just Stop Oil co-authors, both free and in prison.

I think of you behind your prison bars, living in two separate time zones: in real time, today, incarcerated as political prisoners for your beliefs and your deep knowledge of the Climate Emergency; and in future time, tomorrow, living out your 'best life' (as we all will be) on a dislocated and traumatised planet. I hope it's somewhat reassuring for you to know that you will at least be respected, loved and even revered in that future world for what you are doing in real time today. I stand with you in those two time zones, and will continue to do everything I can to amplify your voices and persuade others to honour you as I do.

(Message to Just Stop Oil activists in prison, from the author, 30 October 2024).

JONATHON PORRITT is an eminent writer and campaigner on sustainable development. Since stepping down from Forum for the Future in May 2023, Jonathon has 'returned to his campaigning roots', supporting the Green Party and radical climate campaigns such as Just Stop Oil and Defend Our Juries.

In 1996, he co-founded Forum for the Future, a leading international sustainable development charity, working with business and civil society to accelerate the shift toward a sustainable future. He was also a co-founder of the Prince of Wales's Business and Sustainability Programme. He is President of The Conservation Volunteers and Population Matters, and is involved in the work of many other NGOs and groups.

Jonathon was formerly Co-Chair of the Green Party (1980–83) and Director of Friends of the Earth (1984–90). He stood down as Chair of the UK Sustainable Development Commission in 2009, after nine years providing high-level advice to Government Ministers, and served a ten-year term as Chancellor of Keele University (2012–22). Jonathon was awarded a CBE in January 2000 for services to environmental protection.

His most recent book, *Hope in Hell* (Simon & Schuster, 2020, revised 2021), is a powerful 'call to action' on the Climate Emergency. Before that, he authored nine other books, the first being *Seeing Green* in 1984.

CO-AUTHOR PROFILES
& INTERVIEW EXTRACTS

ALEX DE KONING	**16**, 30, 68, 118/9, 154, 214
ANNA HOLLAND	**19**, 32, 110, 146, 190, 214, 236, 262
AVERY SIMARD	**45**, 56, 138, 154, 220, 236
CHIARA SARTI	**48**, 60, 76, 108, 224, 234
COLE MACDONALD	30, **50**, 84, 140, 154, 182/3
CRESSIE GETHIN	viii, 22, **71**, 90, 166, 192, 246, 262
DANIEL HALL	22, 54, **73**, 90, 138, 156, 240
DANIEL KNORR	98, **99**, 144, 156, 184, 212, 238
EDDIE WHITTINGHAM	viii, 32, **102**, 136, 220, 268
EILIDH MCFADDEN	80, **104**, 108, 140, 216, 234, 262
ELLA WARD	8/9, 62, 92, **123**, 160, 192, 212, 238, 262
EMMA DE SARAM	84, 106, **125**, 144, 158, 210/11, 222
GEORGE SIMONSON	4, 62, 114, **128**, 168, 186, 208
HANAN AMEUR	4, 36, 114, 132, **163**, 240, 260
HARRISON DONNELLY	66, 96, 110, **149**, 164, 184
INDIGO RUMBELOW	26/7, 60, **151**, 186, 199, 216
JACOB PINES	10, 24, 172, **176**, 224, 232
NIAMH LYNCH	10, 22, 54, 168, **178**, 251
OLIVE BURNETT	12, 114, 136, 160, **200**, 224, 260
OLIVER CLEGG	12, 36, 87, 96, **202**, 218, 262
OLLIE SWORDER	66, 184, 196, 218, **228**, 242, 268
PAUL BELL	14, 76, 106, 134, 190, **230**, 238
PHOEBE PLUMMER	14, 40, 92, 134, 246, **252**, 260
ROSA HICKS	42, 170, 196, 208, **254**, 266
SAM HOLLAND	44, 94, 132, 158, 222, 232, 251, **271**
SEAN IRVING	40, 68, 94, 112, **274**

CONTENTS

1: BETRAYAL — 1

2: ON THE FRONT LINE — 23

3: CLIMATE SCIENCE — 53

4: CLIMATE POLITICS — 77

5: THE FOSSIL-FUEL INCUMBENCY — 107

6: EVERYTHING'S CONNECTED — 131

7: TAKING DIRECT ACTION — 155

8: THE WEIGHT OF THE LAW — 181

9: THE EMOTIONAL BURDEN — 205

10: INTERGENERATIONAL JUSTICE — 233

11: WHAT LIES AHEAD? — 257

ACKNOWLEDGEMENTS & CONTACTS — 277

'I first learned about climate change when I was about eight, and had an intense moment of ecological grief when I was nine, kayaking with my mum and brother. For a while, I just couldn't listen to any news about this. Then, as a teenager, I learnt to compartmentalise all this, accepting the world with all its injustice, pain and pollution – I guess I was just like any other teenager!

'But then I read a summary of the "Hothouse Earth" scientific paper, in 2018, when Extinction Rebellion came on the scene, and from then on there was no ambiguity for me about the right thing to do. My first protest was in March 2019, and I never really looked back – although I got a bit disillusioned with XR and chose not to get involved with Insulate Britain, something I now regret. There have been nine arrests with Just Stop Oil since then.'

Eddie Whittingham

'My commitment to justice has steadily deepened from my gap year through to my involvement in Extinction Rebellion and then Just Stop Oil.

'It's always been part of my nature to accept personal responsibility for things once I've seen that there is a problem. I still went through all the rationalisations about how any action leading to arrest was too much risk for someone my age. When I got to uni, I signed up to every climate-focused campaign I could, but it all felt totally disconnected, ineffective and spiritually and physically exhausting.'

Cressie Gethin

1: BETRAYAL

'The world will not be destroyed by those who do evil, but by those who watch them without doing anything themselves.'

Albert Einstein

MY DEDICATION to this book is there primarily to acknowledge, up front, that this book is not some kind of quasi-academic work, looking dispassionately at the phenomenon of Just Stop Oil from different perspectives. It's not remotely objective, though I make what I hope is good use of academic research along the way.

Just Stop Oil 'burst on the scene in a blaze of orange' back in March 2022. Three years on, it announced that it would be bringing to a close all its campaigning activities within the month. During that time, 3,500 Just Stop Oil Supporters were arrested, with around 180 instances of people sentenced or held on remand.

Few, if any, campaigning organisations have achieved such a high profile in such a short period of time, highlighting the impact of today's worsening Climate Emergency, provoking extraordinary levels of controversy for its polarising tactics and attention-grabbing Non-Violent Direct Action.

This uncompromising commitment to civil resistance, as a powerful 'force for change' in the UK, undoubtedly contributed to the Labour Government's decision to end all new oil and gas developments in the North Sea.

I was a supporter of Just Stop Oil from the start. Whatever else one may think about its confrontational tactics, its insistence on 'telling the truth' about the Climate Emergency, exposing the lies and corruption at the heart of politics today, was a constant challenge to mainstream environmentalists and climate campaigners choosing a more consensus-based approach.

My principal connection with Just Stop Oil has been through the twenty-six young activists with whom I've worked to co-create this book. I've interviewed them all, read about them extensively, followed some in detail as their legal processes unfolded, visited some in prison and have got to know some as friends. At the age of seventy-four, I'm fifty years older than most of them, and it's been a privilege to be able to do something at this age that has profoundly changed the way I see the world.

A word about the title of this book. Before opting for *Love, Anger & Betrayal*, my favourite working title was *For the Love of God, Pay Attention!* – as a rather blunt way of capturing the incomprehension (and occasional despair) felt by these young campaigners as we hurtle incontrovertibly towards a world ravaged by climate breakdown, yet very few people seem to care much at all. Because of that, there's an undeniable element of desperation in the choices they make about campaigning tactics, in a world where the old model of political debate seems to be over and 'spectacle beats argument every time'.

As Chris Hayes puts it, somewhat scatologically, in his excellent book, *The Sirens' Call: How Attention Became the World's Most Endangered Resource*:

> What good is persuasion if no one's paying attention? Who cares if people have a negative reaction so long as they have some reaction? You can be polite and civil and ignored, or you can fuck shit up and make people pay attention. Those are the choices in the Hobbesian war of all against all in the attention age, and it's very hard for me to blame these people for choosing the latter.

And please don't think it's only NGOs and activists to whom politicians are not paying attention. In April 2025, the former CEO of Allianz Investment Management, Günther Thallinger, made as dramatic and hard-hitting a contribution to the climate debate (on LinkedIn) as I've heard from any campaigner, pointing out (reasonably and dispassionately) that the world is already approaching temperature levels where insurers will no longer be able to cover many climate risks: 'the math breaks down: the premiums required exceed what people or companies can pay'. At 2°C or 3°C of global heating:

> There is no way to 'adapt' to temperatures beyond human tolerance. That means no more mortgages, so no new real estate development, no long-term investment, no financial stability. The financial sector as we know it ceases to function. And with it capitalism as we know it ceases to be viable.

There was a great article in the *Guardian* about Günther Thallinger's comments, but that was about it by way of mainstream attention. The politicians got on with doing what they had to do coping with a financial world thrown into crisis by Donald Trump's tariff wars.

As with any group of intelligent, highly opinionated, very diverse young people, a little generalisation goes a long way. But here's what I've learned over the past year:

- They do not see themselves as 'brave' or as 'beacons of hope', let alone as some kind of 'latter-day martyrs'.
- Nor do they see themselves as 'eco-zealots' or 'spoilt brats' or 'dangerous extremists'.
- And they seriously dislike the way they are constantly traduced by the UK's right-wing media.

They do see themselves as people who care about science and the critical role that scientific evidence should play in politics. And however flawed democracy may have become in many parts of the world today, they still see it as 'the least worst system' to deliver for citizens today and tomorrow.

Without bitterness or recrimination, they see themselves as having little choice about the decisions they have made, and will acknowledge, though only when prompted, that this entails some inevitable personal sacrifices.

'I'm fortunate as my parents were part of the reason I got into all this in the first place. They're both environmental consultants and I could see it all through their work. It provided the backdrop to a lot of my childhood. I know it's hard for them now, seeing some of the coverage about the actions that we're taking. They approach the crisis we're in from a very different perspective by working within the existing system rather than against it, but they've always been massively supportive of my choices, including the actions I've been part of. This is something I count myself lucky for; I know many colleagues who aren't as fortunate. I can see how it must be any parent's worst nightmare to see their child sent to prison'.

George Simonson

'It's not great having to prepare ourselves for some kind of collapse in the future, but I think we have to recognise that this is now inevitable. Sometimes the only way I can live from day to day is by detaching myself from that future and just getting on with my degree without becoming too anxious.

'This is all balanced out by my sense that a big change is coming! I live in hopeful anticipation, especially when you think about what's already going on today. Whether it's for bad reasons – thinking about what the far right is already doing in many countries – or good reasons – so many people have got behind the Palestinian cause – more and more people are no longer prepared to take it and are getting involved in action. That can also be scary, but complete inertia is much worse!'

Hanan Ameur

Relatively speaking, they are well educated, well supported – not least by the extraordinary community that Just Stop Oil itself has created – living in a country where democracy still functions, more or less, and where basic levels of tolerance, open-mindedness, compassion and empathy are much higher than anything that our toxic media would have people believe. They see themselves as relatively privileged citizens in a world full of pain.

But my generalisations matter little. The principal purpose in writing this book is to allow readers to find out a whole lot more about who these campaigners are, their hopes and fears and why they have chosen to live in 'civil disobedience'. And why all of this is so directly relevant to all of us.

Extracts from the interviews I did with them are included throughout the book, together with some extended contributions. And there is a personal profile for each of them, in their own words.

LOVE, ANGER & BETRAYAL

Astonishingly, the deepest shared emotion these young campaigners gave voice to in their interviews was love: love for all those who have already felt the impact of the consequences of a radically shifting climate – killed, injured, displaced, impoverished, impaired in health, both physically and mentally; love for all those in the future on whom the impact will be greater, and will continue to grow, the majority of whom will be living in countries that have barely contributed to the cause of those impacts – the emission of greenhouse gases predominantly from the rich world and, more recently, China; and love for all non-human life that will be devastated by these worsening impacts. Compassion lies at the heart of their civil resistance.

The 'Anger' of the title is mostly mine, though this is certainly shared by a few of my interviewees. I won't elaborate here. You'll have little difficulty over the course of the ten chapters which follow in discovering for yourself just how abysmal my anger-management skills are.

Which leaves us with 'Betrayal'. On 20 September 2019, inspired by the extraordinary leadership of Greta Thunberg, millions of people, including more than one-and-a-half-million young people, demonstrated in more than 180 countries – the biggest climate protest in history – demanding urgent action from their governments to address the climate crisis. Greta Thunberg herself addressed a gathering of more than 250,000 mostly young people outside the HQ of the United Nations in New York.

But five years on, most of that energy has ebbed away; young people find themselves, yet again, passively observing increasingly dysfunctional political processes, with COP29 in Azerbaijan at the end of 2024 just the latest in a long line of failures. They find themselves stripped of any real agency, able only to beg for the scraps of a so-called 'just transition' that would still leave them to deal with a world devastated by climate breakdown.

This is surely the most abhorrent institutional betrayal of young people there has ever been. Six years ago, attending COP24 in Katowice, Poland, in December 2018, this is how Greta Thunberg put it:

> In the year 2078, I will celebrate my seventy-fifth birthday. If I have children, then maybe they will spend that day with me. Maybe they will ask about you. Maybe they will ask why you didn't do anything while there was still time to act. You say you love your children above all else, and yet you are stealing their future in front of their very eyes.

It's in that context that I used to to urge my friends and colleagues in the mainstream environment movement properly to understand the role of Just Stop Oil, and to be so much more sympathetic to its campaigning tactics, recognising its readiness to shoulder the burden of securing change in a cruelly unchanging world. How is it that these young campaigners seem to find it so much easier – and absolutely necessary – to empathise with future generations than we do, at an age when having children of their own is, for most of them, some way off in the future? Or may never happen. How is it that they feel part of the 'interlinked chain of humanity' so much more powerfully than mainstream environmentalists do?

Many young people today can scarcely believe the way in which the terrifying impacts of climate disruption are proliferating. Yet the vast majority of politicians assert that 'there's nothing to see here' – whether through ignorance, inertia, cowardice or downright self-serving dishonesty and corruption. Just follow the fossil-fuel money that gets them elected and keeps them in power. This dereliction of duty leaves many young climate campaigners both uncomprehending and grief-bound – a cruel form of anticipatory grief as they contemplate the horrors that await hundreds of millions of people through the course of their lifetimes, in a way that can now no longer be avoided.

That kind of grief does not diminish with time – as is the case, one hopes, with the loss of a loved one. It remains at the same level of intensity, day after day without end. Far from being 'the great healer', time becomes the exact opposite, as every day wasted in denial or delay further diminishes our options for mitigating the personal and collective trauma of climate collapse which looms ever larger. Time doesn't heal climate grief. Indeed, it steals the dreams and hopes of all of us, but especially those of the young.

Roman Krznaric uses a quote from Drew Dellinger at the start of his wonderful book *The Good Ancestor*, to which I return in Chapter 10:

> it's 3.23 in the morning
> and I can't sleep
> because my great great grandchildren
> ask me in dreams
> what did you do while the earth was unravelling?
>
> *Drew Dellinger*

With 'climate horizons' getting ever nearer, I suspect angst-ridden parents might soon be unable to get to sleep not because they're worrying about the fate of their great-great-grandchildren in an unravelled world, but about the fate of their own children.

ELLA WARD

Denied the use of any legal defences in her trial, Ella Ward was found guilty of conspiracy to commit public nuisance along with three other Just Stop Oil supporters. This letter was written before their trial, when they had been held on remand for six months.

Hi, it's Ella here, writing from HMP Styal, a women's prison just outside Manchester.

Today is my twenty-second birthday. If you asked me a few years ago what I'd most likely be doing on my twenty-second birthday, I'd probably have guessed eating takeaway and going to the pub. I'd rather be holding a pint of lager and standing in a busy pub than holding a blue plastic prison mug and sitting on my lumpy top bunk!

But being in prison is not a bad way to spend my birthday. Against the odds, I'm doing well, I'm feeling resilient and strong.

Don't get me wrong. I struggle from time to time being in prison. Everyone does. You're stripped of all autonomy, all control, completely reliant on mostly uncaring, overworked, stressed, desensitised prison officers to meet all your basic needs. Prison is degrading and dehumanising. You witness people being treated in a way no human on Earth should ever be.

But all of this being said, if prison is where the government will send me for acting non-violently against harm, and according to my moral convictions, then I feel at peace with being here. Prison is the reaction of a government that feels threatened. People can be shocked that non-violent protest can land you in prison, let alone for up to ten years, which is the maximum sentence I'm facing.

Unfortunately for the government, prison is not a deterrent to living in resistance. They don't understand that refusing to be complicit in allowing the climate crisis to unfold is not a switch we can just turn off. And they don't understand that we act as a community. Each person in prison has a hundred people holding them afloat. With enough community support – the constant stream of letters, visits, money when needed, books, legal guidance and more – we can alleviate the suffering of the people in prison and make it so much more doable.

There is also compassion and care in prison. Every day I see women helping each other out, supporting each other, having each other's backs. When we've got nothing, we're all in the same boat. This isn't to say it's perfect. It's definitely not, but it's also not the scary place full of evil, dangerous people you might have been told it is. So who is benefiting from keeping nearly 4,000 women in prison, at an average cost of £50,000 per person a year? Around nine in ten women are in prison for non-violent offences. In the vast majority, if not all cases, the harm being done by the prison system to those women is far, far greater than any harm these women would do.

These are women worthy of love and respect, kindness and humanity, compassion and care. Women who despite not being treated with any of these things by the prison system, still show it to each other, and show it to me, and have made today – my twenty-second birthday – special and meaningful and warm and memorable. Some women made me beautiful homemade cards, gave me a few homemade gifts, and even pooled a bit of money to buy me two bars of vegan chocolate off the canteen. It was really kind and thoughtful, and I'm so thankful to them.

As the climate crisis worsens, repression of resistance will increase, and the bar for going to prison will get lower and lower as the government continues to try and deter us from taking further effective non-violent action. I refuse to leave my future and the future of young people globally in the hands of a government hell-bent on profiting at the expense of human lives.

I'm completely at the mercy of an uncaring prison system, but I'm surrounded by care from the women in here, and I'm surrounded by care from the people out there.

'I'm a bit different from many others involved in Just Stop Oil, as I never felt that youthful rage and sense of injustice. But I certainly feel it now, about the endless suffering that climate breakdown will cause, today and even more tomorrow. At its simplest, I'm involved because I love life: people and everything on the planet that makes the world worth fighting for.

'History tells us that we just have to accept as activists that some people are going to hate us, trying to raise awareness in this way makes some people very angry. But even when they disagree, there's still a conversation going on. Perhaps they might begin to feel some of that dread?'

Jacob Pines

'I'm a proper ecology nerd, in love with all the life that we still share our planet with. Whether that's wild flowers or birds – thinking about swifts, and whether there will be more or fewer returning to our village. I know it's more likely there will be fewer, but I try to hold onto the possibility that there will be more. Sometimes the grief is too much – past, present and future grief. That's what makes me angry, because I want to do everything I can for the swallows, the swifts, the redwings, whatever. And for everybody we share our planet with.

'If other people can take this on, why can't I? If I can be arrested and am unlikely to be mistreated by the police, why should somebody else have to do that kind of thing for me. As my dad says, it's going to be someone's kid, so why not me?

'It's sometimes scary, but I can't *not* be involved, because that would mean accepting the death of millions of people, birds and animals. I just couldn't live with myself if I accepted that. Every single living being has as much right to life as any other – the buzzard in the sky, my old Labrador, or me.'

Niamh Lynch

INTERGENERATIONAL JUSTICE

Working with young Just Stop Oil activists has forced me to confront the full extent of today's ongoing intergenerational injustice – in effect, the whole notion of Intergenerational Justice has been turned on its head. Instead of older generations doing everything they can to ensure a better, more secure future for all those who come after them, today's younger generation finds itself doing a lot of the heavy lifting to secure a still liveable future not just for themselves, but for their parents and grandparents.

Which is why I hope, almost against all hope, at this very late stage, that all those parents, grandparents and citizens deeply concerned about the future, will find their own way of stepping up.

When Martin Luther King said that 'the arc of the moral universe is long, but it bends towards justice', he sure as hell didn't mean that justice will simply arrive, so sit back and wait for the happy outcome! Such justice is never freely given by those who have power; it is only ever won. To quote Martin Luther King again, 'Social progress never rolls in on the wheels of inevitability. It comes through the tireless efforts and persistent work of dedicated individuals.'

Martin Luther King was particularly concerned, in speech after speech and letter after letter, to remind us that the biggest barrier to progress is not those who openly oppose us, but those who genuinely support the idea of radical change – the end, as it were – without being prepared to engage in or accept the means to that end. As King wrote in his letter written in Birmingham Jail following his arrest for marching in 1963, in response to a critique of the movement by several white clergy members, he wrote, 'Frankly, I have yet to engage in a direct action movement that was "well timed" in the view of those who have not suffered unduly … This "Wait" has almost always meant "Never".'

Please think of these words every time you rush to condemn the tactics of young climate campaigners here in the UK, even as the government and police seek to close them down ever more repressively. And dare I suggest that this applies particularly to mainstream environmentalists?

'Even when I was younger, I felt a lot of anger about how the world seemed so wrong. Friends, teachers and particularly parents just kept telling me not to be so angry – rather than encouraging me into any social justice work! It wasn't until my second year at uni that I started to take action on the climate crisis, with a local group, and then read about Eddie Whittingham's protest at a snooker tournament (see p. 102) which got everybody talking about Just Stop Oil. Okay, I thought, so that's how it works! That led me to go to a JSO talk, where I signed up for slow marching.

'A lot of people are pretending the climate crisis isn't happening, that it's all going to be okay. I know my own involvement can make a lot of friends and family feel uncomfortable – almost as if I am the physical embodiment of all those anxieties! Some friends really don't want to know.

'It's a privilege to be part of this, but it can also be incredibly onerous – doing the work every single day, whether it's front of mind or not, I'm doing it because hundreds of millions of people will suffer in the future. It can be hard being around people who don't understand that – and who don't want to have to think about it.'

Olive Burnett

'Being part of this community really does help. Thinking about the prospect of being arrested would have been pretty daunting on my own – but being surrounded by people who are doing even more radical things made the idea of getting arrested so much more reasonable. Still radical, still disruptive, but much more manageable.

'It's important to me to be part of that. It's a source of pride, however odd that may sound. Who knows what will happen in the future, but at least we'll be able to say to ourselves that we all did our best.'

Oliver Clegg

Back in February 2024, I wrote a blog that began as follows:

> First they came for Just Stop Oil; then they came for radical environmentalists; then they came for Friends of the Earth; then they came for members of the National Trust, the RSPB and WWF. But there was no one left to speak for them.

I know I shouldn't be, but I'm astonished at the apparent lack of visible concern on the part of mainstream environmentalists as we slide inexorably into a police state. The right to peaceful protest remains a basic human right, but you sure as hell wouldn't know that here in the UK any longer.

This point has been powerfully confirmed by Michel Forst, the UN Special Rapporteur on Environmental Defenders under the Aarhus Convention, when he issued a report after his visit to the UK in January 2024. In his own words:

- Peaceful protesters are being prosecuted and convicted under the Police, Crime, Sentencing and Courts Act 2022, for the criminal offence of 'public nuisance', which is now punishable by up to 10 years imprisonment. The Public Order Act 2023 is also being used to further criminalise peaceful protest.
- In some recent cases, presiding judges have forbidden environmental defenders from explaining to the jury what their motivation is for participating in a given protest, or from mentioning climate change at all.
- Prior to these legislative developments, it had been almost unheard of since the 1930s for members of the public to be imprisoned for peaceful protest in the UK. I am therefore seriously concerned by these regressive new laws.

I return to these matters in Chapter 7.

'I started with all the standard campaigning stuff – lobbying my MP and council – but it really was like bashing one's head against a brick wall! So then I got involved in an Extinction Rebellion group in Birmingham, where I first learned about direct action – I even got myself accidentally arrested when drumming with the XR Samba Band!

'And then I went to some JSO talks, which really rammed home the urgency of the situation – as well as coming up with solid solutions. Since then, I've been arrested eight more times, although most weren't actually followed up on. In November 2022, I climbed a gantry on the M25, for which I was sentenced to twenty-two months in prison in August 2024.

'When I was in prison, there was this amazing support network, co-ordinated by Rebels in Prison Support, that made it all possible to deal with. No one can be truly resilient on their own; you need people around you who understand what you're going through. This kind of support started with XR, but the JSO community really is extraordinary.'

Paul Bell

'With my adoptive family, I found a much more loving acceptance of seeing the world for what it is, though this wasn't easy. I'd grown up hearing that any climate crisis was a very distant threat, that there was nothing particularly to worry about as world leaders were already 'on it'. But in 2019, I read one of those big reports from the IPCC, and it was as if my world was falling down around me! It made me realise that all that stuff about banning plastic straws, turning the lights out, saving the polar bears, signing petitions and going on official marches was little more than co-ordinated gaslighting.'

Phoebe Plummer

1: BETRAYAL

The windows of the Ministry of Justice, which is responsible for these new laws, display all sorts of historical statements and aphorisms, intended no doubt to show how deeply those inside care about justice. One of them is John Locke's famous saying from 1689: 'wherever law ends, tyranny begins'. But what if the laws by which you as a citizen are bound are themselves tyrannous? In other words, using power or authority in a cruel and oppressive way. That is exactly what is happening here in the UK today.

No one has covered this astonishing erosion of some of the most basic rights of UK citizens better than George Monbiot in his *Guardian* column:

> Why do the mass killers of the fossil fuel industry walk free while the heroes trying to stop them are imprisoned? ... Why, when we know so much, do we permit a handful of billionaires to propel us towards predictable catastrophe?
>
> Such questions invite trouble. Those who raise them are either sidelined or, if they cannot be ignored, relentlessly attacked. It is so much easier to lock up the people impeding our frenetic dance towards oblivion and then pretend the problem has gone away.

When the maximum sentence for chaining yourself to railings is more than twice the maximum sentence for racially aggravated assault, anyone who cares about justice should be appalled. The draconian new laws introduced by the Conservatives, which, shockingly, the Labour government now defends, ensure that non-violent protest is routinely treated as a more serious crime than most forms of violence.

It isn't just young people today who are betrayed by what is going on here in the UK. It is every single one of us.

ALEX DE KONING

I'm a Scottish, twenty-six-year-old PhD student in green hydrogen production. I have been a spokesperson for Just Stop Oil since the summer of 2022. In my spare time, I am a magician, a wannabe screenwriter, and a self-proclaimed movie buff.

INVOLVEMENT

I never went to any kind of protest, including marches, before joining Just Stop Oil.

Two weeks after my first JSO talk, I helped block an Esso oil terminal in Birmingham on 1 April 2022; I didn't get arrested.

I got arrested for the first time for blocking a bridge with XR, then for sitting on top of an oil tanker with JSO, and then for occupying oil pipes in Clydebank for forty hours.

I took seven months out of my PhD to do JSO full time. Got arrested for helping two people glue onto a frame in Kelvingrove Art Gallery. I was the mobilisation co-ordinator for Newcastle, working with Anna Holland, as well as training to be a spokesperson. My first proper interview was in August 2022 with Jeremy Kyle, on Talk TV, and now I've lost count of how many I've done.

In October 2022, I became a 'spokes-fixer', helping to set up interviews for others, and a 'spokes-trainer'. My fifth and final arrest to date was in November 2023 for slow marching. Despite five arrests, I have only one conviction, for helping others to glue onto the frame in Kelvingrove.

PERSONAL PROFILES

MOTIVATION

Over the course of my activism journey, my motivation has evolved. At the beginning, it was more of a sense of 'this is the right thing to do', and I want to be on the right side of history when climate activism eventually wins. It was exciting; I was meeting great new people, and I liked the feeling that I was doing something that really mattered. Now, it's more of a feeling that if I don't do this, who else will? A feeling of I should use the privilege that I have because there are so many people around the world who cannot do activism in the way that I can and who are suffering from the emissions that countries like mine are continuing to emit. It feels like much more of a civic duty that I should be doing.

INSPIRATION

Honestly, seeing first hand how hard some of my friends in JSO work under constant stress, or how calmly they are willing to sign up to prison, or get back on the horse when something goes wrong, is what inspires me the most to try harder in my activism. Spokespeople like myself and Anna are often interviewed and shown to the public, but some of the people behind the scenes are a million times more worthy.

IN NATURE

Loch Lomond. It has sentimental value to me because I've camped there several times, including with Anna right before we started dating and then again, a year and a half into our relationship on a perfect holiday. It's also very close to Glasgow, where I grew up, and it always amazes me how close such beautiful nature is to such a big industrial city, and how easily accessible it is by train!

QUOTATION

The future isn't cast into one inevitable course. On the contrary, we could cause the sixth great mass extinction event in Earth's history, or we could create a prosperous civilization, sustainable over the long haul. Either is possible, starting from now.

Kim Stanley Robinson

RESOURCES
Finite: The Climate of Change, a film by Rich Felgate.
On Fire, Naomi Klein's book, is equally fantastic.

WHAT LIES AHEAD?
Success in my eyes would be a complete remodelling of the economy. With no stock market, billionaires, pressures of economic growth etc. it will be a lot easier for people to move away from cities, split into small communities, slow down their daily life, focus far less on careers and more on how they can help the community such as by growing food, creating citizens' assemblies, spreading education etc.

I became one of the spokespeople for Just Stop Oil in August 2022 – a role that can be very challenging. Apart from all the major climate-induced disasters, the impacts of the climate crisis can be much harder for people to see. I've heard commentators talk about the phenomenon of 'enshitification', with so many things becoming slightly more shit all the time! Insurance costs going up in many different sectors. Food prices increasing. Crops harder to grow because of uncertainty in weather. Disruption to travel ...

The biggest challenge is how to explain that the climate crisis is linked to so many other things. When Just Stop Oil organised an action in October 2023 to block a bus taking refugees to the Bibby Stockholm, people just couldn't understand what that had to do with new oil and gas licences in the North Sea! What most people are focused on is the cost of living, asking why we're not more actively involved in addressing that. Which we are, as it happens, by talking constantly about high energy bills, dependence on imported gas, vulnerability to global price setting, and so on.

(See pp. 118–119 for an account of Alex's action at Clydebank).

ANNA HOLLAND

I'm twenty-two years old and I'm a poetry student. Aside from causing trouble to the system, I fill my days with poems. I'm always writing, it's how I make sense of this broken world that has given me a broken heart. I've always loved the outdoors and go camping every chance I can get. My friends and family are the most important things in my life. Everything I do, I do for them.

INVOLVEMENT

I first got involved in climate action in 2019 during the school strikes. From then I worked with Youth Strike for Climate Manchester, Climate Live UK and the Youth Climate Justice Coalition until early 2021. I turned my attention to the anti-democratic, anti-protest Police Bill and joined Kill the Bill, organising marches across Newcastle. In May 2022, I went to a Just Stop Oil talk held at my university, and that led me to work full time as a mobiliser for the movement. My first arrest was in August 2022 for blocking an oil tanker outside Kingsbury Oil Terminal. I was arrested twice in October 2022 for blocking roads in London, and then again for throwing soup at Van Gogh's Sunflowers. Since then, I've been working in outreach and fundraising and as a spokesperson for Just Stop Oil.

I was sentenced to twenty months for the Van Gogh action on 27 September 2024, and released on Home Detention Curfew on 27 January 2025.

MOTIVATION
I can't keep watching the news, watching the collapse of everything in real time, and doing nothing about it. The leaders of the world seem intent on destroying everything we hold dear for the sake of dirty money. I need them to know that I will fight them every step of the way. They won't succeed, and they will be held to account for what they have done. I want to be there for that day.

INSPIRATION
It's not a real person I attribute my drive to, but a fictional one. Alba Trueba, from Isabel Allende's The House of the Spirits, *inspired me not just to be revolutionary but to be resilient. She, ironically, drove a car with a sunflower painted on the door, so that the rebels fighting for revolution in Chile knew that she was part of the resistance. She was imprisoned and tortured, but she never lost her purpose. Reading this story at the age of fifteen was fundamental to building the person that I am today. When I had bad days in prison, I thought of her, she gave me strength.*

IN NATURE
The Welsh coast calls me. At night I hear a steady rumble in my ears, and when it gets louder, I know it's time to return. The smell, not of salt but of the universe, fills my nerves even now. Even when I am miles away. When I visit, the rocks press warm against my feet like kisses welcoming me home. I know there is not a single thing in this world that separates us. Every part of me is made up of the coast. My lungs breathe in and out like the tide. My hair rests against me like seaweed. My teeth are pearls and my nails are shells. My eyes are glistening rocks, glowing in the sun. I miss it every day.

QUOTATION
As a queer, female-bodied person, my resistance is rooted in a deep tradition of those who have stood on the front line of protest. With that in mind, Audre Lorde's line 'we were never meant to survive', from her poem 'A Litany for Survival', is one that I always return to.

RESOURCES
There is a book and documentary that I would like to recommend: Irina Ratushinskaya's memoir, titled Grey Is the Colour of Hope, is a fantastically inspiring book that shows the power of non-violence and non-compliance even in the harshest of regimes. Also, Rich Felgate's documentary Finite: The Climate of Change is a powerful tale of small communities making a fantastic difference, and it really drove me to step over the boundary from passive to active resistance.

WHAT LIES AHEAD?
The success of the climate movement is something I dream about often. Issues that are both caused by and have caused the climate crisis are intrinsic to so many of the issues that pervade our society. The climate crisis and the cost-of-living crisis are both crises of greed. If we solve the problem of greed, of stark individualism that has been perpetuated by the capitalist machine, we can solve so much. Success to me looks like small communities. The 'us' and 'we' rather than the 'I' and 'me'. It is renewable energy powering human society. A reduction of the scale and sheer globalisation of our lives. It is a system based on equity, honesty and safety above all else. This kind of world is being rejected by such a small percentage of people. We can beat them. We can win the life we deserve.

(See p. 146 for two of Anna's poems written from prison.)

I really didn't get Extinction Rebellion to begin with – it sort of weirded me out, with all the fun and games, music, colour and so on. I'd actually been deeply depressed getting to grips with the implications of the climate crisis and was really grieving at the time. A real mismatch.

It was different with Insulate Britain, where I just felt the emotion was much more raw, that they were more like me as a person, and I was completely blown away by their courage. They spoke to ordinary people in a way that hadn't happened before. So, it was a pivotal moment for me. We realised then that we would have to change our lives to reflect that. So, we got rid of the mortgage, I extricated myself from the family business, and worked out how to balance bringing up two young children with the sort of activism which we wanted to commit ourselves to.

Daniel Hall

'When I was younger, I was involved in Fridays for Future, the school strikes, which I helped organise in Bedford – making banners and posters! But it achieved nothing, or at least not enough, given how little time we have.

'Getting involved in Just Stop Oil wasn't so much a "why" as a "when". There are other ways of doing this – getting into policy and so on – but there's no time for that. We need to be doing things RIGHT NOW! I don't understand why everybody isn't running around screaming at what's happening, why people aren't really, really panicked! If my house were on fire, I wouldn't simply watch it burn. And our house *is* on fire! Our beautiful planet, our home, is burning – we need to save it – and to save ourselves.'

Niamh

2: ON THE FRONT LINE

'A great deal of intelligence can be invested in ignorance when the need for illusion is deep.'

Saul Bellow

THIS IS a book about one particular group – young Just Stop Oil activists – within the wider Just Stop Oil organisation, which was itself one part of the so-called 'Radical Flank' of campaigning climate non-governmental organisations (NGOs) in the UK, which in turn make up one part of a very broad climate movement. This is a mature and socially progressive ecosystem, all too often characterised – and then dismissed – as an undifferentiated 'green blob' by its critics and opponents. The Radical Flank in the UK is part of an international movement that constantly ebbs and flows. At the time of writing, the A22 Network represents similar organisations in around a dozen countries.

Just Stop Oil set out to be a polarising organisation from the start, saying, 'We are not here to please.' Its spokespeople went to great lengths to explain that its use of Non-Violent Direct Action (NVDA) was fully justified, both tactically and morally, given the Climate Emergency that has been officially declared by the UK Government. It rebuffed any suggestion that its actions were over the top, as is sometimes asserted because the UK itself is only a small contributor to global greenhouse gas emissions (about 1% of the total, although 3% when looked at cumulatively since the start of the Industrial Revolution), and is widely seen to be one of a small cohort of leading countries when it comes to emissions reduction.

The UK's emissions have gone down by around 52% since 1990,

'There's always going to be conflict in my mind about public disruption. The whole point of an action is that it's meant to be disruptive – that's what the media are interested in, what gets people talking about the crisis. And the economic impact of the disruption puts direct pressure on the government. But there's always a question about how to achieve that impact whilst minimising harm and striving for healing as part of the process. We had conversations about all this; I knew that it was a source of creative tension, that it was part of making the decision to do it.'

Cressie

'The really frustrating people are *not* those who continue to deny that the climate crisis is real, or even those whose vested interests mean they want to see people kept in the dark. It's those well-meaning liberal types who claim to be really concerned about the crisis, but not if there's any inconvenience involved – just as Martin Luther King said! Then, they suddenly find themselves defending the status quo, and are totally opposed to people even walking in the road to protest. But I don't get too angry. It takes people a long time to get to the point where they're willing to stand up against incumbent forces in our society – it took me a long time!

'I hear a lot from people saying our tactics are counterproductive – even from friends and family! Shouldn't we be trying to do it differently, without disrupting people, without making so many people angry? And shouldn't we be doing everything we can to keep the politicians on board, given the gravity of the climate crisis?

'I'm not sure that people have properly thought this through. I like to flip a discussion when I'm accused of being misguided: 'Tell me exactly what you think I should be doing, and why you think it would be so much more effective?' It's not as if we've made much progress over past decades and it doesn't look like things will change any time soon.'

Jacob

2: ON THE FRONT LINE

significantly more than most countries, primarily as a consequence of decisions taken under Margaret Thatcher back in the 1980s to start using gas rather than coal for generating electricity. In September 2024, the UK became the first major industrialised economy to put a complete end to using coal to generate electricity. Coal still accounted for 40% of the UK's electricity as recently as 2012, since when renewables have gone from strength to strength – from 7% in 2012 to 50% in 2024.

The other undisputed source of climate leadership in the UK is the 2008 Climate Change Act, the first such cross-party legislation anywhere in the world – and the absolute bedrock of UK efforts to transition to a low-carbon economy. An independent Committee on Climate Change was established under the Act, and it's remained genuinely independent since then, often causing Ministers significant embarrassment in its highlighting of fairly consistent under-performance.

This relatively progressive backdrop has given direct action campaigners plenty to work with, and both Extinction Rebellion (XR) and Just Stop Oil can claim to have had real success with their NVDA-based campaigns. After XR's so-called 'Easter Rebellion' in April 2019, the UK Parliament was the first in the world to declare a Climate Emergency a month later. And there is no doubt that XR had a marked impact on the Conservative Government's decision in June 2019 to amend the Climate Change Act by committing to 100% net-zero emissions by 2050. It backed that up by pledging that all of the UK's electricity would come from non-fossil fuels by 2035. The Labour Government is pursuing an even more ambitious target of 2030.

On top of that, Just Stop Oil's principal demand – that the UK Government should issue no new licences for any oil and gas investment – became Labour Party policy while it was still in opposition.

Those I interviewed for this book give only 'grudging recognition' of some of the claims for UK leadership, rapidly followed by reminders that it's 'all relative' – as in relative to what the science is telling us and to the scale of the Climate Emergency.

INDIGO RUMBELOW

EARLY DAYS

Anyone who has been involved in campaigning about climate change here in the UK thinks back to that time when there was a sudden explosion of interest. The movement was already there, but it was disparate and fragmented. Then, in 2018, it ignited with the School Strikes, the incredible leadership of Greta Thunberg, and the emergence of Extinction Rebellion. There was a real sense that things could change, and we were seeing some really game-changing results. I began to see the real power of civil resistance, of people from across society coming together, and the ability of ordinary people to disrupt the capital city with demonstrations.

After the protests in April 2019, representatives of XR met with Michael Gove and Parliament and hundreds of local authorities declared 'Climate Emergencies' of one kind or another. People were starting to get serious about long-term net-zero targets and the government commissioned a Citizens' Assembly. All three of XR's demands were met, at least in part.

INSULATE BRITAIN

Insulating British homes is a no-brainer policy. It brings people out of fuel poverty, whilst dramatically reducing the emissions from energy use, a win both for people and the climate. Insulate Britain's road-blocking tactics sparked fury across the political spectrum, but we weren't there to be liked. Time and again we heard the words, 'We agree with what you're saying, we just don't like how you're saying it!' Support for our demand grew even while the campaign was berated in the press. We seeded the debate in public and political consciousness, and two years later the Tory Government brought in the Great British Insulation Scheme. Though current policy falls far short of what needs to be done, we were pleased to have got the issue out there, and that other campaign groups are continuing to push the demands forward.

JUST STOP OIL'S ORIGINS

Just as Insulate Britain grew out of XR, so Just Stop Oil grew out of Insulate Britain. We wanted to shift the focus back onto fossil fuels directly. In February 2022, we wrote to Boris Johnson demanding a stop to all new licences for fossil fuels in the North Sea. Basically, we were told to sod off! So in March, we went public with a series of protests at oil terminals, petrol stations and cultural events.

Just Stop Oil was able to mobilise a higher percentage of young people at that time than Insulate Britain. A lot of Insulate Britain activists carried over into Just Stop Oil – it was pretty much full-time for many of us by then. The energy then was really intense and exciting, although it was emotionally and physically exhausting.

Just Stop Oil decided to hang up the high-viz in April 2025: this is an extract from Indigo's commentary after that decision, written while on remand in HMP Styal:

> After three years, Just Stop Oil is ending its campaign of non-violent civil disruption. Fifteen Just Stop Oil supporters are currently locked up for refusing to obey governments whose climate inaction is frankly murderous. Continuing with oil and gas for decades is an act of violence that will unfold over the coming years; it is the ultimate betrayal of the people.
>
> In a world where it is legal – and profitable – to extract and burn oil, gas and coal, it is unsurprising that attempts to disrupt the system have led to imprisonment. Supporters of Just Stop Oil have court trials listed well into 2027. The State continues to put in the dock the students, doctors, vicars, scientists, teachers, the ordinary people who support Just Stop Oil, while letting the fossil fuel bosses off the hook. It must abandon these show trials and invest these resources into investigating the real criminals.
>
> It does feel bittersweet, that this chapter of the climate movement is closing while I am sitting here in prison. But it is our duty to adapt our resistance to this new reality, and to follow Mahatma Gandhi's concept of 'Satyagraha': 'non-co-operation with evil is as much a duty as co-operation with good'. It is time to consider a new design for a mass movement capable of confronting the many intersecting crises that we face. As we head to the drawing board. I invite you to join us.

2024: THE HOTTEST YEAR EVER

The whole of the next chapter is devoted to the science of climate change. It's impossible to exaggerate the importance of this to radical climate campaigners: the more radical, the more important respect for and attention to the science becomes. This is the primary source of justification for taking controversial and often confrontational decisions about appropriate actions.

In that regard, the situation at the end of 2024 couldn't possibly have been more disturbing:

- The average temperature in 2024 was 1.6°C above pre-industrial levels, making it not just the hottest year ever, but also breaching for the first time the totemic 1.5°C threshold that has dominated the debate over the last ten years;

- The majority of scientists now believe that our chances of staying below 1.5°C are 'deader than a doornail', as renowned climate scientist James Hansen put it;

- The concentration of CO_2 in the atmosphere, as recorded at Mauna Loa in Hawaii, increased by 3.58 parts per million (ppm) to 427 ppm (1 ppm is equivalent to 1 milligram of a substance in a litre of water), the biggest annual jump since records began there in 1958;

- Wildfires around the world worsened this record increase, on top of emissions from burning coal, oil and gas – all of which also increased in 2024;

- Both the Greenland and Antarctic ice sheets lost record volumes of ice in 2024, and average ocean surface temperatures broke all records, month after month.

2: ON THE FRONT LINE

I used to watch Just Stop Oil spokespeople grinding their teeth in frustration at having to explain the science of climate change on media outlets like GB News – constantly interrupted and attacked for acting as messengers on behalf of the global science community! Behind all the contested statistics and topical debating points there's still 'the big stuff' – basic things like the Earth's Energy Imbalance, with more and more of the incoming energy from the sun not being reflected back into the atmosphere, but trapped by the greenhouse effect – that 'blanket' of heat-trapping gases like CO_2 and methane, slowly destabilising the systems that have made life on Earth so productive and so resilient. The problem is really very clear: as long as more energy is coming in than is getting out, the planet will continue to get hotter.

It's not easy to keep that big picture in mind. Indeed, the 'whole truth' about climate change can be unbearable. Our brains are simply not wired in that way – and we're all susceptible to what is called 'psychic numbing'. For instance, when scientists explain that this energy imbalance is the equivalent of the heat generated by 500,000 Hiroshima bombs every day – that's five bombs a second – those little synapses just begin to shut down! It's much easier to put one's faith in those political leaders who keep promising some kind of magical technology-driven energy transition – even as fossil-fuel consumption keeps on rising!

In my experience, Just Stop Oil activists were always much more realistic than most environmentalists in gauging just how significant some of today's technological 'breakthroughs' really are in practice. If we look at solar energy, for instance, we have to hold two insights in mind at the same time:

1. The solar revolution is already well underway, and it will indeed transform global energy markets within the next five years. No less an authority than *The Economist* magazine, in the article 'The Exponential Growth of Solar Power Will Change the World' in its special issue in June 2024, explained the nature of this revolution: 'an energy source which gets cheaper the more you use it

'I don't get annoyed that most young people aren't involved. So many young people are just tired, with no money, nothing to fall back on, no security. The reason I didn't get involved myself was quite simply because I had nothing.

'When I first heard about Extinction Rebellion, I really couldn't relate – in 2018, it meant nothing to me. That lasted quite a long time, even when Just Stop Oil came on the scene – I didn't like them at the start either! I really didn't understand how throwing soup at a painting could make any difference. Then, in October 2023, I went to a big demonstration in London, 'Oily Money Out', and met a lot of people from Just Stop Oil who turned out to be completely different from how I'd imagined!'

Cole Macdonald

'I've met some of the loveliest people, and some of the most knowledgeable, both about the history of social change and social-change theory. We talk all the time about our 'theory of change'! At one level, it's completely crazy to think that people disrupting traffic or an airport will lead directly to a change in government policy – we all understand that isn't quite how it works. But what does work is making sure that the climate crisis stays as visible as possible, that the media can't just go on ignoring it. We have to constantly reinvent ways to get that kind of media attention. Spraying private jets, orange cornflour on Stonehenge or throwing soup at a Van Gogh painting – people may not like these things, but they keep the climate crisis in the news.

'So little has changed over the last decade or more that we have to keep on disrupting that complacency, trying to force people to think about the true nature of the climate crisis.'

Alex de Koning

marks a turning point in industrial history'. This is indeed huge: 'To call solar power's rise "exponential" is not hyperbole, but a statement of fact. Installed solar capacity doubles roughly every three years. The next ten-fold increase will be the equivalent of multiplying the world's entire fleet of nuclear reactors by eight in less time than it typically takes to build a single one of them.'

2. So, all good? Absolutely, and the momentum is unstoppable. But the solar revolution does NOT give us a 'get out of jail free' card, and will do little, on its own, to reduce the risk of runaway climate change. It will NOT remove the stranglehold that the fossil-fuel industry has on the global economy (as I explain in Chapter 5). Moreover, this kind of breakthrough will NOT easily be repeated, and certainly not with Electric Vehicles. And it will NOT, on its own, ensure that emissions of greenhouse gases start coming down at the very steep rate that is now required.

Just Stop Oil activists understand this as a matter of course, not because they're gloom-ridden misanthropes intent on crushing whatever spark of hope we latch onto, but because both things are equally true!

Here's a much more attractive variant of 'psychic numbing'! In 2004, it took a year to install a single gigawatt (GW) of solar power; in 2010, it took a month; and in 2016, a week. In 2024, a single day – with prices falling all the time, and correspondingly huge positive implications for poor and developing countries.

'A lot of people get upset that this is quite confrontational and think we should be more focused on winning people's hearts and minds. I think this is ridiculous. We can't have public perception as the benchmark for whether something works or doesn't – I don't think that would have helped the Suffragettes! Even if people hate Just Stop Oil, it doesn't mean they aren't agreeing with us that we need much stronger action. That's the so-called 'Radical Flank effect', as I understand it.

'What annoys me are not the ignorant or self-absorbed, or those in denial, but the 25 per cent of people who are broadly on board, but very complacent about what is now needed. They keep asking all the wrong questions and stick with their reformists' paradigm even though it isn't working! They are often highly educated, with considerable privilege, but stick with tokenistic campaigning, like marching, completely trapped within an 'ideology of progress' with its implicit assumption that things go on getting better over time. There's absolutely no acknowledgement that civilisations collapse completely from time to time!'

Eddie

'I've always been involved in climate campaigning – marches, petitions, countless letters to my MP, who I'm pretty sure is sick of me! I was very involved in the 'Kill the Bill' campaign in 2020/2021, fighting back against the Police, Crime, Courts and Sentencing Act. A million people signed the petition. But we lost that fight, and the movement just seemed to die out.

'That was my first experience of facing a big loss. I realised that marches and petitions can be ignored so easily, and they *are* ignored all the time. I didn't know how to continue the fight, and felt helpless – as if the whole world was falling apart. That's why I got involved in the Newcastle Just Stop Oil group. It was like finally resurfacing after being underwater for a year. It reinvigorated me.'

Anna Holland

2: ON THE FRONT LINE

TOXIC POSITIVITY

That's the truth of it, and it doesn't make for easy reading. There's a huge army of 'climate solutionists' out there trying to cajole and coerce us all into 'staying positive' and 'not giving way to dangerous despair', which basically means not sharing 'the whole truth' of what we're up against, but only as much of the truth as we think people can cope with. When are we going to learn that it's simply not possible to protect people from the pain of the crisis that has been inflicted on us?

And there's something else about all this 'toxic positivity', as some people describe it. When I'm listening to 'keep the faith' protestations of hope from politicians and business leaders, I get the impression that what they really want us to keep faith with is today's economic orthodoxy – in all its planet-trashing, obscenely wasteful and deeply inequitable and cruel manifestations. But is that really the kind of 'hope' we should be advocating for, tethering us to Western lifestyles, consumerist fantasies and the same kind of entitled 'manifest destiny' ideology that got us into such a fix in the first place?

This is the point at which we need to get stuck into the debate about tactics in the climate movement.

Just Stop Oil epitomised what is referred to as the 'Radical Flank' of the environmental movement. It organises disruptive actions specifically designed to polarise opinion, and to encourage the 'still undecided' to decide which side they are on. This contrasts with a rather more amorphous Moderate Flank, relying on more conventional, non-controversial tactics to bring about incremental change.

People tend to overlook the fact that the reason why the Radical Flank is radical is not just because it has a different view about tactics and how to change the status quo, but because it has a much more radical view of politics in general – and, in particular, about the life-destroying stranglehold of neoliberal capitalism; about the chronic limitations of today's representative democracies; about the utter absurdity of thinking of progress in terms of exponential year-on-year economic growth;

and about the profound immorality of today's grotesque disparities in wealth.

In other words, the Radical Flank comes with a whole lot of radical baggage! Some climate campaigners see that as a good – and totally necessary – thing; others see it as a massive misstep given that any transformation of today's fossil-fuel-intensive economy can only be secured through the support of 'majorities of people, not marginalised minorities'. That approach necessarily entails some studious de-politicisation, just as it tends to be associated with a communications strategy which shares some of the truth about climate change, but definitely not the whole truth!

THE RADICAL-FLANK EFFECT

This is an important debate, and one with a tremendously rich historical lineage. Taking a closer look at the two social movements most often cited by Just Stop Oil activists as inspirational analogies for their own movement – namely, the Suffragettes and the Civil Rights Movement in the USA – the idea of a 'Radical Flank effect' is a particularly important hypothesis: that a small number of people using confrontational, unpopular and often illegal tactics create additional space for more moderate campaigners to make the case for change from within the system.

Many people know the story of how the Suffragettes, through the Women's Suffrage and Political Union (WSPU) of Emmeline Pankhurst, split away from the mainstream suffragist movement, the National Union of Women's Suffrage Societies (NUWSS), in 1903. But fewer people are aware that the Suffragettes themselves split, in 1912, when members of the Union were forced to choose between the Union's original non-violent militancy and the increasingly violent tactics of Emmeline Pankhurst and others as they engaged in a 'guerilla war' with the government of the day. This has huge ramifications for Just Stop Oil today, with its very strict emphasis on the discipline of non-violence.

2: ON THE FRONT LINE

One can see the same phenomenon with the Civil Rights Movement. In 1963, Martin Luther King and the Southern Christian Leadership Conference positioned themselves as the Radical Flank to the much more conservative National Association for the Advancement of Colored People, which was supported by large numbers of white liberals. But historians point out that the Southern Christian Leadership Conference had its own Radical Flank, represented by Malcolm X, Kathleen Cleaver, Angela Davis, the Black Panthers and the Nation of Islam. In his famous letter written from his cell in Birmingham Jail, Martin Luther King urged established politicians to tolerate the non-violent protest tactics of his own organisation in case 'millions of Negroes, out of frustration and despair, will seek solace and security in black nationalist ideologies', and claimed that without his organisation's philosophy of non-violence, 'by now many streets of the South would be flowing with floods of blood'.

As we'll see in Chapter 7, both Emmeline Pankhurst and Martin Luther King became less and less tolerant of the mainstream movements from which they had broken away. Pankhurst cursorily dismissed those who claimed the Suffragettes were alienating lifelong supporters of female suffrage by asking them: 'What did your sympathy do for us, my good friend, when we had it? It is better to have you angry than to have you pleased, because sooner or later you will come to the conclusion that this intolerable nuisance must be put an end to' – and the fact that a majority of MPs expressed their personal support was dismissed with equal gusto: 'we had reached the stage at which the mere sympathy of members of Parliament, however sincerely felt, was no longer of the slightest use'.

Neither Pankhurst nor King, nor the movements they led, were there to be liked. As the climate movement's Radical Flank, Just Stop Oil was not there to be liked either. And its activists asked the same questions of the mainstream environment movement: 'What has your sympathy for the cause achieved in practice?'

The truth is, sadly, that we're looking back over more than thirty years of near complete failure – from the emergence of the United Nations

'I can't believe how quickly people have forgotten that Just Stop Oil has already had a massive impact, persuading Labour not to issue any new oil or gas licences. Whenever somebody tells me we're just annoying people without having any impact, I want to shout out to them to remind them what has actually happened!

'It feels good to have been a small part of that, helping to build momentum by keeping the climate crisis in people's minds. That definitely happened with our action at a performance of *Les Misérables*, which generated a huge amount of coverage. (We were found guilty of aggravated trespass, and I got a sentence of eighty hours of community service.)

'It was extraordinary to see how the previous government tried to crack down on all protests of this kind. And it seems to be no different under Labour! In a funny kind of way, however, that kind of repression rather proves our point.'

Hanan

'My first role with Just Stop Oil was as a mobiliser, organising meetings and giving talks. It turned out that I was really, really bad at that – and the meetings were pretty depressing! I'd be talking about closing down the fossil-fuel infrastructure in a few months' time, and the two or three people in the room would say, 'You and whose army?' A lot of students had bought into the myth that once you've been arrested, your whole future – jobs, security – would be irretrievably ruined.

'There was definitely a lot of over-optimism at that time within Just Stop Oil. I can remember people talking about mobilising up to 10,000 students! That figure kept coming down, and that was when we began to think about a different way of doing things – with far fewer people. We had no option but to try something different – pitch invasions and disrupting cultural events.'

Oliver

Framework Convention on Climate Change in 1992 to the crushing cumulative data at the end of 2024. Compared to 2023, emissions of greenhouse gases were up; concentrations of CO_2 in the atmosphere were up; levels of subsidies for fossil fuels were up; new investments in fossil fuels were up; and the average temperature increase during 2024 was up, since the time of the Industrial Revolution, by a record 1.6°C.

Even though these are global outcomes, rather than being specific to the UK, we cannot disregard them. Besides, even in UK terms, we know we have failed more than we have succeeded, a failure in which I and many others share, for having relied on a theory of change – gradual, consensus-driven incremental change, which will eventually deliver the stable climate on which we all depend – that now stands undeniably exposed.

But did the polarising, much more confrontational theory of change espoused by Just Stop Oil prove to be any more effective? The honest answer to that is that I don't know.

I've been struck by just how many of my colleagues in the Green Movement did not feel positive about Just Stop Oil and were worried about what might be described as the 'negative Radical Flank effect', or 'backfire effect', with more people being alienated from the cause, rather than attracted to it. An important paper, 'Extreme Protest Tactics Reduce Support for Climate Movement', published in January 2025 by social scientists from New York University (NYU) and Aix-Marseille University (AMU) argues exactly that point.

By contrast, a study carried out by the Social Change Lab back in December 2022 tracked the responses of around 1,400 members of the general public before and after a week-long campaign by Just Stop Oil to block the M25, specifically asking them if they would be more or less likely to support Friends of the Earth (not Just Stop Oil itself) as a consequence of that campaign. The percentage increased from 50.3% to 52.9% of respondents saying they would. That's a 2.6% gain – the equivalent of 1.75 million people in the UK.

The Climate Emergency Fund (one of the biggest funders of NVDA campaigns around the world) points out that most claims of a negative Radical Flank-effect are based on responses from interviewees asked to consider hypothetical situations, 'not from studying effects in the real world'. Rather than a substantive backfire effect, what we were probably looking at here was an all too familiar 'second-hand *Daily Mail* effect'. (I'm assuming that my colleagues are not actually reading the *Daily Mail* themselves!)

This debate obviously mattered to all those I've interviewed for this book. Some were wholly comfortable with the 'not here to be liked' positioning; for others, the case that most of the arguments used to attack Just Stop Oil from a tactical perspective (its tactics are ineffective, it alienates otherwise sympathetic people, it damages the reputation of the wider movement etc.) didn't really amount to very much, is reassuring.

MORAL CONVICTIONS

However, this whole 'theory of change' debate ignores the deep moral convictions that drive all those I've interviewed for this book. It's these convictions, not tactical calculations, that provide the deeper rationale for prioritising NVDA, for choosing civil resistance over other campaigning tactics. Those moral commitments also made it easier for them to cope with the fact that the UK has become one of the most authoritarian democracies in the world with regard to undermining the right to protest and to freedom of speech. Research by the University of Bristol in November 2024 showed that the UK arrests peaceful protesters three times as often as the global average in democracies.

When the trial of the 'Whole Truth Five' came to court in July 2024, Michel Forst, the UN Special Rapporteur for Environmental Defenders, was there to witness proceedings. He was particularly concerned about one of the defendants, Daniel Shaw, who received a four-year sentence:

> I fail to see how exposing Mr Shaw to a multi-year prison sentence for being on a Zoom call that discussed the organisation of a peaceful environmental protest is either reasonable or

proportionate, [or] pursues a legitimate public purpose. Rather, I am gravely concerned that a sanction of this magnitude is purely punitive and repressive.

The same clearly applies just as much to the other defendants: Roger Hallam, who got five years; and Louise Lancaster, Lucia Whittaker De Abreu and Cressie Gethin, who all got four years. And for all five, Forst commented that 'defendants should be allowed to explain why they have decided to use non-conventional but yet peaceful forms of action like civil disobedience when they engage in environmental protest'.

(In March 2025, these sentences were reduced in the Court of Appeal. Lady Justice Carr described the five-year sentence for Roger Hallam as 'manifestly excessive', reducing it to four years. Shaw and Lancaster's sentences were reduced to three years, while Whittaker De Abreu's and Gethin's sentences were reduced to thirty months.)

It's hard not to see these legal constraints and sanctions as anything other than politically motivated. It's hard not to see the judiciary in this country as having been fundamentally compromised by this politicisation. And it's hard not to see those who have been incarcerated as anything other than political prisoners in terms of the sacrifices they have made. These views are widely shared by human rights campaigners, including Liberty, Amnesty and Climate Rights International, which has done detailed research into the impact of the new anti-protest legislation brought in by the former Conservative Government – covered in detail in Chapter 8.

Even the right-wing *Sun* newspaper was somewhat outraged back in March 2025 when twenty uniformed officers from the Metropolitan Police broke open the door of a Quaker Meeting House to arrest six young women who had hired a room to discuss their concerns about the climate crisis and the ongoing genocide in Gaza. They were detained on 'suspicion of conspiracy to cause a public nuisance', and their homes were subsequently raided to 'secure additional evidence'. As the Recording Clerk for the Quakers said:

'I went to my first Just Stop Oil talk in 2022. I was pretty sceptical, hearing statistics about the rate of species extinctions and realising how inaccurate they were! I've had reservations about JSO tactics since then. Its 'urgency culture' results in a lot of broken activists, which probably isn't an effective long-term strategy. There might be a moral justification for this, but treating activists as expendable can't be sustainable.

Its basic theory of change is fine – providing a deeply unpopular, headline-grabbing radical flank that helps the ecological crisis become more present in people's minds and makes them more sympathetic to moderate organisations – but I'm not sure the organisation really knows what to do beyond that. I can't see that asking government to make further changes will work without far greater numbers involved in more disruptive actions. So, I don't feel the kind of loyalty to Just Stop Oil that some others do.

Sean Irving

'I was fascinated by Extinction Rebellion and Insulate Britain, but still not certain that this was for me. I still wanted to go to university, and a bit of me was thinking about the career ladder and having a family. Like most people that age, I wasn't certain that I wanted to give up on those dreams and had a hundred good reasons why it should be other people who needed to step up! But then I signed up with Just Stop Oil, and it went pretty quickly from seeing just how important this was – and many of the things we hope for in the future are just lies – to seeing it as the *most* important thing. That's a big shift, and it really does change what one thinks of as "normal".'

Phoebe

2: ON THE FRONT LINE

No one has been arrested in a Quaker Meeting House in living memory. This aggressive violation of our place of worship and the forceful removal of young people holding a protest group meeting clearly shows what happens when a society criminalises protest.

The whole idea of 'sacrifice' is extremely sensitive amongst Just Stop Oil campaigners. The various manuals laying out both the theory and practice of Non-Violent Direct Action – and there are many! – all stress the importance of three essential elements:

DISRUPTION 'Polarisation is necessary'; 'conflict is the very heartbeat of social movements'; 'trying to avoid negative reactions from people by being less disruptive is folly'.

ESCALATION 'Without scale, there can be no lasting transformation'.

SACRIFICE Perhaps most succinctly captured through 'Deeds, not Words', the watchword of the Suffragettes. With hundreds undergoing unspeakable suffering through 'forcible feeding' as they persisted with their hunger strike.

I found that all those I interviewed were reluctant to get drawn into extended conversations about sacrifice. I can't recall a single occasion when the word was used without my prompting. And there was a very high level of self-awareness about this – an appreciation that causing disruption that breaks the law without being prepared to pay the legal price, does not play well with the public. It risks turning bystanders against the cause, rather than building sympathy. Indeed, if the price is high, as in disproportionately harsh sentences, there is an acknowledgement that this might help build a reservoir of greater sympathy. Academics refer to this as 'the paradox of repression'.

> 'I have a lot of compassion for people who clearly care about the climate crisis but aren't willing to risk arrest at the moment. It took me such a long time. I'm very conscious that I haven't faced the barriers to involvement that many others do – I don't have to hold down a demanding job, pay off a mortgage, worry about looking after somebody else or deal with difficult health issues. In Manchester and Australia – and now in London – I've always had supportive friends; we all felt part of a community, talking about the same concerns. I know this is all part of the privilege of being able to get engaged. It's simply not practical for many people to make that commitment.'
>
> Rosa Hicks

However, others are more than happy to do the talking about sacrifice, which is not hard to do. This is how Naomi Klein, inspirational climate campaigner and author, responded to some of the excessive sentences imposed on Just Stop Oil activists in 2024:

> In a world that was right-side up, you would be celebrated as the ones who helped break the spell that is setting our world on fire. In truth, your actions could still do that, if enough people know about them. I think our elites do understand this threat that your actions represent. It's the threat of mass uprising. And that threat is precisely why ethical direct action in defence of planetary life is being criminalised so viciously in the UK and in many other countries. But I know you could not have anticipated these draconian sentences. Please know that, although it will feel otherwise at times, you are not alone. You and your actions will not be forgotten. I am convinced that your brave choices will impact in the world in ways we cannot yet anticipate or even imagine.

These words powerfully capture what it feels like for Just Stop Oil's young campaigners, especially now that Just Stop Oil itself has shut up shop. This is a horizon that stretches out not just for a few years into the future, but over many decades – as recently highlighted by a quite extraordinary report from the Institute and Faculty of Actuaries (IFoA), the UK's chartered professional body dedicated to the profession.

Published in January 2025, 'Planetary Solvency – Finding Our Balance with Nature', in partnership with scientists at the University of Exeter, robustly critiques orthodox economic predictions which estimate that the impact of an average temperature increase of 3°C by the end of the century would be around 2% of annual gross domestic product (GDP). 'These estimates are precisely wrong,' the report reads, 'rather than being roughly right, and do not recognise there is a risk of ruin.' The Institute's risk management experts diligently reassessed risks associated with impacts such as fires, flooding, droughts, temperature increases and rising sea levels through to 2050 and on to the end of the century.

As we'll see in the next chapter, there is now a very strong likelihood that we'll experience an average temperature increase of at least 2°C by 2050 – an outcome described by the report's authors as 'catastrophic'. Take a deep breath and get your head around the projected impacts associated with that 2°C rise:

- Economic contraction – GDP loss of over 25%
- Mass human mortality events resulting in over 2 billion deaths
- Warming of 2°C or more triggering a high number of climate tipping points
- Breakdown of some critical ecosystem services and earth systems
- Major extinction events in multiple geographies
- Ocean circulation severely impacted
- Severe socio-political fragmentation in many regions
- Loss of low-lying regions
- Mass migration of millions driven by heat and water stress
- Catastrophic mortality events from disease, malnutrition, thirst and conflict

Two billion deaths by 2050. That's just twenty-five years away. And for the final kicker, bearing in mind that we're currently on a business-as-usual trajectory towards at least a 3.7°C temperature increase by 2100, the contraction in GDP would rise to 50% and the number of projected deaths to 4 billion.

Insights as shocking as these are as hard to process whether you hear them from world-class scientists, famously conservative and cautious actuaries or supposedly radical organisations such as Just Stop Oil. But we need to grasp them.

FOR THE LOVE OF GOD, PAY ATTENTION!

'We live in a country where one third of kids are living in some form of poverty, with more than a million in destitution – and we're the sixth largest economy in the world! With Labour, we have another government that refuses to address chronic inequality, promising instead to deliver the same old economic growth that simply benefits the elites. I feel real rage when I think about the evil that these elites are perpetuating, both in Palestine and with the climate. Can anybody really be surprised that there is a lot of anger and rage amongst campaigners for Palestine after all these decades – a rage that sometimes means that the police are seen as the intermediate enemy which has to be taken on.

'Whilst the vast majority of young people have not taken any action to stop the elites, I don't believe it is because of apathy. No young person wants to inherit a future of economic collapse, starvation, social breakdown. All I know, as far as young people are concerned, is that the rage will grow and grow over the coming decades, because things are going to get worse and worse.'

Sam Holland

AVERY SIMARD

I'm a twenty-three-year-old computer engineering graduate from Canada. Since graduating, I moved to London, and am now working as an activist full time, the likelihood of which, even one year ago, would have seemed like a far-fetched dream. I grew up in Calgary in Alberta, a city built off of, and desperately attempting to hang onto, the oil and gas industry. I managed to escape Calgary to go to university in Ontario, and from there, got the chance to do an internship in London as part of my degree. It was over that year-long internship that I got involved with Just Stop Oil and found that activism is where my true motivation lies.

INVOLVEMENT

I first got involved with Just Stop Oil in March of last year, and haven't stopped since then. I went back to Canada for nine months to finish my degree in September 2024, and while there, got involved with Just Stop Oil's sister campaign in Canada, Last Generation Canada. Since coming back to London in June of this year, I've been working full time with Just Stop Oil and volunteering with Youth Demand. The only action I have taken so far is slow marching with Just Stop Oil last summer and swarming with Youth Demand this fall.

MOTIVATION

It's a mixture of immense rage, love and inspiration. It seems that the strongest and most consistent emotion I feel is rage – rage about the endless injustices of the world, the systemic inequality, the denial and wilful ignorance of politicians and media. This simmering rage keeps me going, but it grows from the seeds of love. I find myself in frequent awe of life on this planet, and love for the people around me, and the things I am able to experience.

INSPIRATION
When I first moved away from home, I printed out photos of Nadya Tolokonnikova, of Pussy Riot, and Assata Shakur to stick on my wall. I remember seeing Pussy Riot's infamous protest at Moscow's Cathedral of Christ the Saviour on TV when I was eleven years old. I didn't understand what was happening at the time, but that image stuck with me.

And Assata's autobiography was truly radicalising. They both represent ferociously radical women, who have both faced extreme forms of repression, and constantly inspire me to continue in this work.

IN NATURE
I can't choose just one, but I feel most at home on Hampstead Heath in London and in the Rocky Mountains in Alberta, Canada. My heart and my life are split between these two places, with my family in Alberta, but most of my friends and current life based in London. The Rockies will always hold a special place in my heart. There is something unbelievably striking about standing in the midst of such huge, natural formations. It provides a feeling of your true scale both in time and in size. As for Hampstead Heath, it is my favourite slice of nature in my favourite city in the world, a constant escape to nature in the time that I've lived here.

QUOTATION
> We have chosen each other
> and the edge of each other's battles
> the war is the same
> if we lose
> someday women's blood will congeal

upon a dead planet
if we win there is no telling
we seek beyond history
for new and more possible meeting.

Audre Lord, from 'Outlines'

RESOURCES

The Reality Bubble by Ziya Tong is a book that I will endlessly recommend. While it doesn't necessarily provide concrete pathways to action, it frames the climate crisis in a really unique way by looking at humanity's greatest blind spots. She frames the reader's existence as a sort of cosmic joke, that the likelihood of you being alive at this time, in the midst of this crisis, is infinitesimally small. But that means you have both the unbelievable opportunity and responsibility to resist, to be a part of the change that is so desperately needed. Nothing has inspired me to take action on the climate crisis as much as that book did.

WHAT LIES AHEAD

I don't think the fight will be over in just a few years, I believe that civil resistance will likely be necessary for the remainder of my lifetime and probably far beyond – if humanity still exists at that point!

However, in the next few years, we will have won if governments around the world sign the Fossil Fuel Non-Proliferation Treaty and actually follow through with their promises in doing so. We will have won if we've successfully completed a full, just transition to renewable energy.

My dreams of 'success' extend far beyond those few, necessary demands though. If we want to prevent the continuation of the climate crisis, we need to address it at its root, at the systemic level. We need to overhaul our current capitalist economic system and create a truly democratic revolution to reform our broken political system. If we want to address injustice and create truly positive change, we need to begin striving towards participatory democracy and a more communal way of living.

(See p. 56 for Avery's reflections on Canada's wildfires in 2024.)

CHIARA SARTI

I'm a twenty-five-year-old mathematician, forced to face up to the reality that doing puzzles as the house burns down is psycho behaviour. I've been radicalised beyond all imagining by the daily humiliation of trying to access transgender healthcare in the Britain the Tories made. Driven by the renewable energy of civil resistance against the British regime for their crimes against humanity. Hobbies include cooking, biking, scheming and mischief.

INVOLVEMENT
Involved with Just Stop Oil, and co-founded Youth Demand in January 2024. Four arrests with Just Stop Oil and one with Youth Demand. I was jailed for nineteen days for marching down a road for just twenty minutes – obviously against the wishes of the British state.

MOTIVATION
Unless we fight for our lives, we will all die. Like it or not, we have the massive responsibility of resisting the biggest injustice in human history. If we sit on our collective arses and pretend we don't have an obligation to do everything we can to limit total catastrophe, the next one thousand generations will spit on our graves. Better to sleep in prison beds than to stay up at night losing our minds over the obscenity we're witnessing.

PERSONAL PROFILES

INSPIRATION
Larry Kramer, founder of Gay Men's Health Collective and ACT UP. He took shit from absolutely no one, and saved countless people, one blunt truth at a time. Many people I love would be dead without his tireless resistance.

IN NATURE
Grantchester Meadows in Cambridge. It's so gorgeous it even has a Pink Floyd song dedicated to it:

> See the splashing of the kingfisher
> Flashing to the water
> And a river of green is sliding unseen ...

QUOTATION
'How do you keep on fighting when everything is lost? Ask a Palestinian. A Palestinian is someone who is wading knee-deep in rubble. Palestinian politics is always already post-apocalyptic: it is about surviving after the end of the world and, in the best case, salvaging something out of all that has been lost.'
Andreas Malm, The Walls of the Tank, 2017

RESOURCES
The Nutmeg's Curse, by Amitav Ghosh.

WHAT LIES AHEAD?
We have front-row tickets to witness fossil capitalism destroy itself! This is a thing to celebrate in itself. Sure, we may all starve to death, but amid all the grimness, we will have stood together, shoulder to shoulder, and created a thing of beauty. We will have danced on the grave of the old regime, and no one can take that away from us.

COLE MACDONALD

I was born in 2002. I'm a 'raging queer feminist' with a history of student protest, and an interest in the intertwined nature of the law and political ideology.

INVOLVEMENT

I'm active in anti-genocide, anti-apartheid protest, and was arrested for the first time on 2 May 2024 while blockading the L3Harris arms factory in Brighton. We successfully blocked the dispatch of bomb-release mechanisms. I am also active in the squatting scene.

I was twenty-two at the time of my arrest and imprisonment with Just Stop Oil, having been arrested at Stansted Airport for criminal damage – spraying private jets with orange paint – interference with key national infrastructure and aggravated trespass.

While awaiting trial, I am on a deferred year at uni, writing my dissertation, 'Nomadic Identity as the Most Radical Form of Anti-Capitalism', in which I delve into the roots of capitalism through colonialism and land privatisation. Outside of direct action, I like writing – poems and songs, mainly – making zines, which are usually about police surveillance or 'know your rights' stuff, camping and being out of doors. I sing songs of resistance with my comrades, my favourite being a re-written version of 'I Wish There Was No Prisons'.

MOTIVATION

My motivation for taking action is a genuine belief that it will work. I am taking action in full awareness of my power and the empowerment of others. My pride and my courage empower me to take action. As silly as it sounds, when I'm terrified before an action, which I always am, it is my own bravery and the awareness of my hope that empowers me to take action. It is knowing that many people before me have done this and done more and lost more and gained more because of it that keeps me going. There is no other path in my life that I would have been happy taking other than this one.

INSPIRATION

There are many people around me who inspire me constantly, and they hold my hope. My squat crew inspire me a lot with their community and their love and their ability to write songs in any situation. My friend Luke Elson, who is currently imprisoned in Wormwood Scrubs, inspires me a lot. We discuss queerness and feminism on the phone – their ability to live so fully in prison inspires me. Phoebe Plummer and Rosa Hicks, my most hilarious and hopeful friends, find everything funny, even the most mundane things ever. Phoebe finds hope and solace in everything, and Rosa brings joy everywhere. My gorgeous friend Sylvia Mann has been a huge inspiration in facing down my upcoming prison sentence and has been a rock of hope.

I think the idea that we are not just individuals, but a sea of undivided hope is the biggest inspiration. Truly knowing and truly feeling that what we do is right and true and brave is the biggest inspiration of all.

IN NATURE

I went to the Lake District just before I ended up in prison. This was also my first solo holiday ever, so that was pretty exciting. But to be honest, anywhere in nature is home to me. The first thing we did when we were allowed out – our first time outside after three days of solitary confinement – was sit on the grass. Jen (with whom I did the action) dug her hands into the grass and looked for bugs to remind her that other living things were here.

QUOTATION

When I was in a police cell, in solitary confinement for three days, I was having a pretty rubbish time, and a saying just came to me:

> I am strong
> I am brave
> I care
> I am here for a reason
> These spaces are built to make you lose hope
>   and be passionless
> But by having hope
> And being passionate
> I am resisting.

The first thing I did when I got to prison was write it down and stick it to my wall with toothpaste, and I read it every day and repeated it every day. I used it as an affirmation and a meditation. And to this day, when it all feels a bit too much, I do the same.

RESOURCES
This book!!!!!! (Thank you, Cole!) Also, anything by Noam Chomsky or Angela Davis. If reading is not your thing, there are accessible speeches online that are also hugely inspirational.

If you want to take action against the climate crisis, don't limit your research to climate resistance, reach out and learn about all sorts of resistance movements and political theories. We cannot overcome this issue without the learnings and inspiration of our comrades who went before.

WHAT LIES AHEAD?
OMG! What a huuuuge question! To put it pretty simply: no cops, no prisons, no fascists!

(See p. 182 for Cole's account of three days in solitary confinement.)

3: CLIMATE SCIENCE

THE FOLLOWING question was put to the Just Stop Oil contributors to this book: 'Why are most politicians still so indifferent to scientists' warnings about the climate crisis?'

They responded as follows:

- 'It's just baffling. Totally baffling.'
- 'They still think all the bad stuff is decades away – which means they've got plenty of time to get it sorted!'
- 'Well, I don't want to be rude, but they're only indifferent because they're ignorant.'
- 'I'm not sure they know what's going on – *really* going on, not superficially.'
- 'Not many of them are scientists, are they? I wonder if they ever get to talk to climate scientists?'
- 'I think they hate having to deal with the climate crisis. It forces them to make trade-offs – big trade-offs – between the present and the future.'
- 'I don't know. Simple as that: I just don't know.'
- 'Well, they can see "our house is on fire" – as Greta Thunberg put it – but I guess they think the fire brigade is coming from somewhere else?'
- 'A friend of mine puts it down to a lack of imagination – just not possible to grasp the scale of the monsters we're shaping.'

I was very struck by this collective sense of bemusement amongst

'We got into growing food with a small market garden – a kind of 'off-site activism'! In a very small way, we're reflecting the massive transformation we need in today's unsustainable, exploitative food system. Food is at the heart of so much that matters to people, and when we connect people to the process of growing that food, deeper change becomes possible. We see it very much as a front-line job, building up real solidarity, working on the basis of mutual aid – emphasising those instincts that neoliberalism has beaten out of us. It's part of the adaptation we're going to need in a very different and climate-stressed world.

'But it's absolutely not "deep adaptation" – running for the hills and preparing for the apocalypse! Real survivalist stuff! It's the exact opposite, giving people a chance to get their hands dirty and to enjoy the connection between growing and eating good, healthy food.'

Daniel H.

'The way we live should be aligned with our goals. You shouldn't get this out of proportion, however – having a one-minute shower rather than a five-minute shower is not going to save the planet! But I still couldn't live with myself if I had a five-minute shower! Living aligned to my values gives me the strength to keep fighting.

'And it's the same thing with flying. Lots of people say not flying is just stupid, as the plane is going to be flying anyway. But for me, not being on that plane is part of my activism. It's about standing up to a business-as-usual status quo. I know I'm fortunate being able to make decisions of this kind, but we're all part of a much bigger system, and what we choose to do or not to do is an important part of that system.'

Niamh

3: CLIMATE SCIENCE

this small group. I'd characterise their collective view as follows: 'Why doesn't the science speak for itself? And speak to them as disturbingly as it does to us?' And the fact that it clearly doesn't, otherwise we'd be well into 'emergency response mode' instead of 'steady as she goes', leads to bemusement, consternation and serious questions about what happens to democracy if citizens can't rely on policy being shaped by such an overwhelming scientific consensus.

It's not that climate change is unique in this regard. The slow but inexorable drift away from 'evidence-based policy' can be seen in many other areas – the impact of ultra-processed food on people's health, for example, or the need to rethink drugs policy, or how to deal with plastic pollution.

But in few, if any, of these areas is the gap greater between 'what the science is telling us' and the policy response to that consensus. And in no other area are we forced to confront that this gap is what lies between a viable, secure future for humankind and the terrifying breakdown of human civilisation.

And yes, it really is that stark. In my conversations with campaigners, it was precisely that implacable dichotomy – act now, or else it all just collapses – that underpins so much of their bemusement.

And that's because they are paying attention to what's happening on the front line. Back in March, I was rung up by one of my co-authors to ask me what I thought about a video on YouTube ('Immense Methane Leaks in Antarctica: A Hidden Climate Threat Unveiled'), picking up on an article in *Polar Journal* detailing how a Spanish research project had observed huge quantities of methane being released from the seabed around the Antarctic Peninsula. I knew absolutely nothing about it! So I got Googling, having written a lot in the past about methane leaks at the other end of the world in the Arctic – and about the possibility of massive methane pulses. (Methane is a greenhouse gas many times more potent than CO_2 at trapping heat in the atmosphere.) It was impossible not to be shocked: at my own ignorance; at the potentially devastating consequences of this phenomenon; and the fact it caused not a ripple outside the elite world of polar scientists.

AVERY SIMARD

Our primitive brains can barely imagine the ferocity and scale of many of today's wildfires. Proximity to such primal phenomena leaves people scarred in very different ways, but can also leave people more polarised than they were before, rather than more united in their grief – as explained by Avery Simard:

> On 22 July 2024, a wildfire was reported near the town of Jasper in Alberta – a town situated in the national park of the same name, and a well-loved place for all who have the privilege to experience it.
>
> By 25 July, 25,000 people had been evacuated from the town, the fire had grown to a size of 32,000 hectares, and a twenty-four-year old firefighter had died. When the fire finally died down, Canadians saw images of unbelievable devastation.
>
> While wildfires have become something of the norm in Canada during the summer months, for such a cherished place as Jasper to succumb seemed to break through the usual rhetoric and appeal directly to people's emotions. The Premier of Alberta, Danielle Smith, was fighting back tears as she spoke in a press conference, saying, 'Although those of us who experience Jasper as visitors can't imagine what it feels like to be a Jasperite right now, we share the sense of loss with all of those who live in the town, who care for it, and who have helped build it. Jasper, we will continue to stand by you.'
>
> While she seemed truly genuine in her emotions, climate campaigners found her words not only hollow, but even insidious. Since coming into office, Danielle Smith had led a concerted anti-renewable energy campaign, citing vague 'environmental protection' reasons, and imposing a seven-month suspension on all new renewable energy projects in the Province. Alberta produces eighty-four percent of Canada's oil. Smith talks constantly of 'tripling down on Alberta's role in the world as an energy superpower – and we will not apologise for that.'
>
> The image of Danielle Smith in that press conference, crying over the destruction in Jasper, is a horrible reminder of our broken system – one that will consistently place profits over life, seemingly no matter the cost.

3: CLIMATE SCIENCE

Climate activists are often accused of acting 'irrationally' or 'emotionally'. Yet, in my experience, I've rarely come across a group of people so exercised about the importance of science – and about the 'scientific method' itself.

This is Wikipedia's definition of the scientific method:

> The scientific method is an empirical method for acquiring knowledge that has been referred to since at least the seventeenth century. It involves careful observation coupled with rigorous scepticism, because cognitive assumptions can distort the interpretation of the observation. Scientific enquiry includes creating a testable hypothesis through inductive reasoning, testing it through experiments and statistical analysis, and adjusting or discarding the hypothesis based on the results.

And that's exactly what more than forty years of climate science has been all about, involving tens of thousands of scientists in dozens of countries, covering multiple facets of 'why' and 'how' the climate is changing – presided over by the Intergovernmental Panel on Climate Change (IPCC), set up by governments in 1988 specifically to advise them on the science of climate change. It lays down the consensus-based line in great door-stopper reports every five or six years. So well was it judged to have carried out that remit that in 2007 it was awarded the Nobel Peace Prize, alongside former Vice President Al Gore.

Just Stop Oil activists know all this, even though they're not huge admirers of the IPCC itself. They see it as being far too slow, always several years behind the reality of what's happening on the front line of our changing climate, and far too susceptible to brutish political pressure from the fossil-fuel incumbency (see Chapter 5). But if you're looking for an institutional manifestation of the scientific method in practice, then the IPCC is as good a gold standard as you're going to get.

Outside of the IPCC process, there are countless initiatives involving scientists from many different disciplines, either contributing research in their own specialist areas or offering multi-disciplinary overviews. One of the first of these was 'World Scientists' Warning to Humanity' written by Henry W. Kendall, a former chair of the board of directors of the Union of Concerned Scientists, way back in 1992 – before any of the interviewees for this book were born. It opened with the blunt warning that 'Human beings and the natural world are on a collision course.'

Fast forward thirty-two years to last year's 'State of the Climate' report with its chilling conclusion: 'We are on the brink of an irreversible climate disaster. This is a global emergency beyond any doubt. Much of the very fabric of life on Earth is imperilled.'

This kind of report is standard fare for many Just Stop Oil activists – although quite a few of those I interviewed acknowledged that they avoid excessive 'doomscrolling' to help keep themselves sane! But they're never far away from that scenario of potential 'societal collapse', that literally 'existential' moment for humankind, and the report's considered use of the word 'irreversible' will have sent a few additional shivers down their backbones.

It's not really the out-and-out climate denialists still in our midst – including the current President of the United States, who continues to describe climate change as a 'hoax' – who really worry them. They're more concerned by those who read the same scientific reports that they do, but seem to take nothing from them. It appears that their continuing indifference is mainly down to 'climate illiteracy' at every level of our political system.

So, on behalf of the twenty-six young people whose insights have shaped this book, let's dig down into this a bit. In the absence of any serious media pressure, and in the relative absence of what is still referred to as 'postbag pressure' – constituents demanding better of their elected representatives in large numbers – what's really going on? There are still five underlying reasons for this collective political failure.

3: CLIMATE SCIENCE

1. POISONOUS DOUBT

As soon as the IPCC was set up in 1988, the fossil-fuel industry realised that its worst enemy would be the science itself – not least because of all the work they themselves had done more than a decade earlier. So they promptly set up the Global Climate Coalition in 1989, and from that point on, enthusiastically supported by the American Petroleum Institute and countless agents of the fossil-fuel incumbency, billions of dollars have been invested in sowing doubt: questioning any emerging consensus; critiquing individual studies as they emerged; attacking the reputation of individual climate scientists and their institutes or university departments; constantly disparaging the work of the IPCC itself; casting doubt on direct links between emissions from fossil fuels and warming in the climate; feeding misleading stories to ideologically-friendly editors, journalists and commentators; giving tacit support to utterly unscientific nonsense about sunspot activity and the like; and paying unscrupulous scientists to peddle sceptical and overtly hostile 'studies'. On and on it went, year after year; drop after poisonous drop into the ears of ill-informed politicians and the body politic as a whole.

Over the years, as a result, there have been many scientists keen to undersell the likely impacts of climate change rather than oversell them. Whenever cracks appear in the usually rock-solid consensus amongst climate scientists, politicians are quick to exploit those differences, to propagate their own variety of 'delayers' doubt', while, for the most part, remaining keen to avoid being tainted with the 'denialist' label.

I've devoted a whole chapter to this further on. The cumulative damage done over those thirty-five years is beyond comprehension, let alone financial calculation. It's all been done knowingly by those in positions of power, however 'unknowing' the tens of millions of people employed in these industries might have been. It's left an intellectual and cultural legacy as toxic as the physical pollution and devastation their facilities and infrastructure have caused all around the world. This is industrial force majeure at its most brutal.

'Right now, we should be particularly worried about the whole food chain. If there's a reduction in availability of crops of 20 per cent, that doesn't mean that prices will rise by just 20 per cent. Until now, global markets have been able to manage food security crises, but we simply don't have the experience of farmers in a country like Ghana devastated by the climate crisis.

'People don't seem to understand that societies can – and do – collapse under extreme pressure. The closest we have come to that recently in a rich country like the UK is when people were fighting over toilet rolls at the start of the Covid pandemic! People were talking about how supply chains could fall apart in a couple of weeks.'

Chiara Sarti

'There is so much disinformation about climate change – are things are as bad as some people say they are or is there nothing to worry about? Who should younger people particularly be listening to? When Just Stop Oil emerged in 2022, Sir David King, the former Chief Scientific Adviser to the UK Government, was making strong statements that we had only three or four years to put in place the radical decarbonisation necessary to ensure a stable climate in the future. And yet the editors of the *Daily Mail* and *Daily Telegraph* tell their readers, day after day, that this kind of authority is not worth listening to. There are millions of people in the UK today who are completely disenfranchised by not knowing what is going on, by the approach of the Government and our predominantly right-wing media.

'What has happened to the enthusiasm of all those kids who went on the school strikes back in 2018? There seems to be so much disinformation these days that taking action gets harder. I'm sure media misrepresentation has a lot to do with it.'

Indigo Rumbelow

2. COMFORT BLANKETS

Most politicians at least appear to recognise that the climate crisis is 'big and scary'. So they often seize on comfort blankets of one kind or another to keep that fear at bay. The ridiculous, and largely unscientific, obsession with 'not exceeding an average 1.5°C temperature increase by the end of the century' provides a classic case. When this ambition was inserted into the final communiqué of the Paris Conference in 2015, it was seized on and then endlessly repeated by politicians as if to demonstrate just how serious they were about addressing the crisis. It duly became subject to the kind of rhetorical abuse in which politicians like Boris Johnson are so fluent.

Scientists keep on trying to point out that this 1.5°C figure was then, and still is, little more than symbolic, and that the laws of physics tell us that in reality there is no guarantee that the warming will halt even if we succeed in dramatically reducing concentrations of CO_2 in the atmosphere. By that stage it's possible – likely, even – that we will have triggered any number of feedback loops in natural systems: melting of the permafrost and sea ice; tree dieback; loss of peat bogs, which will drive future warming regardless of the cuts in emissions that we make.

This is just one part of a wider phenomenon where setting targets becomes a substitute for making real cuts in emissions, though some countries are more serious about making emissions cuts than others.

The EU has a target of cutting emissions of greenhouse gases by 55% by 2030, and that target, together with Putin's invasion of Ukraine, has undoubtedly helped to drive real performance improvements. Emissions fell by 8% in 2023 as more coal-fired plants closed down and investment in renewables increased. Emissions are now 37% below their levels in 1990 across the EU – making the goal of reaching 55% by 2030 hard, but not unachievable.

'I can really empathise with people my age who don't see things the same way I do – even though we're all doing the same environmental science course, and learning exactly the same things about the likelihood of climate breakdown! Not all that long ago, I was in their shoes, and hadn't connected emotionally with what I was studying. But I'm surprised that I'm the only person on my uni course who is trying to do anything about it, and I don't just mean getting arrested, but even doing more conventional stuff, painting banners with Extinction Rebellion, going on a nature march! A lot of people took this course precisely because they cared more about the climate crisis than most of their friends, but now I guess they're going to become environmental consultants and sort out all those green tick boxes on some company's sustainability register. Because that's what people with environmental science degrees do!'

Ella Ward

'It was in 2022 – the year we had the 40°C heatwave – that I started to hear about Extinction Rebellion. I was quite critical, as an engineer, and wasn't convinced that the situation was as bad as people made it out to be. But then I started doing my own research and discovered that it really *was* that bad. And it rapidly dawned on me that this wasn't a scientific or technological problem – it's a political problem. We've got so much technology already. Why aren't the politicians taking advantage of all that? I was considering a career in engineering or making furniture, but thinking about all those tipping points in the future – food shortages, climate disasters and mass violence – led directly to my first action, gluing myself to the road in Parliament Square in October 2022.'

George

3. COMING BACK FROM 'OVERSHOOT'

Some comfort blankets are much bigger than others, but the biggest of them all is the collective fantasy that we could still get back on a net-zero trajectory later in the century even if we do 'overshoot' by failing to deliver on emission reduction targets over the next two decades.

This mad idea of overshoot was written into the IPCC's scenarios right from the start, but it began to gain real momentum after the signing of the Kyoto Protocol in 1997. It was explicitly designed to provide more breathing space and to let governments off the hook in terms of the really steep decarbonisation process that they would otherwise have been obliged to sign up to.

This is such a bizarre idea for most people that I really need to paraphrase the IPCC's logic, which goes something like this:

> We are very unlikely to achieve the kind of accelerated decarbonisation that we really need, which means we will inevitably end up with dangerously high concentrations of greenhouse gases in the atmosphere.
>
> This will exacerbate levels of warming, which in turn will drive dire climate impacts.
>
> But, don't panic: despite that regrettable overshoot, it will still be possible for us to draw billions of tonnes of CO_2 out of the atmosphere in the future, eventually bringing concentrations of greenhouse gases back down to a level where a stable climate becomes available to us again.

Even 'bizarre' doesn't quite cut it: 'insane' would be a more accurate way of describing this plan, especially when you hear how we are planning to do it: either through engineered solutions, the current poster child for which is a technology known as Direct Air Capture, or through nature-based solutions, such as recarbonising our soils; protecting and extending our forests, wetlands and peat bogs; and restoring mangrove swamps and sea-grass and kelp beds.

Right from the start, most scientists were deeply sceptical that it would be possible to come back from overshoot. But the alternative would have been to tell governments that if they wanted to avoid runaway climate change, they had no option but to drive down emissions by at least 7% or 8% each year. And that was deemed 'politically impossible'. So the politicians gratefully wrapped that great big fur-lined overshoot blanket around themselves, allowing emissions to go on rising every year – as they continued to do even in 2024 – reassured by the prospect of a booming withdrawals industry to deal with the overshoot at some point in the future.

We now know the full extent of this deception. Investors have already poured hundreds of millions of dollars into Direct Air Capture technologies over the past decade. It sort of works, technically, but the volumes of carbon being withdrawn from the atmosphere are negligible, and the cost per tonne is beyond exorbitant.

Nature-based solutions offer more realistic options for removing CO_2 from the atmosphere, and I'm an enthusiastic supporter of massive investments, starting right now, to recarbonise our soils, protect and extend our forests, wetlands and peat bogs; also, in the sea, to restore mangrove swamps and sea-grass and kelp beds. I would even support investment in Direct Sea Capture – there are much higher concentrations of CO_2 in the oceans than there are in the atmosphere. These are genuinely exciting opportunities, with a real potential for 'gigatonne impacts', as in billions of tonnes re-sequestered in natural systems, year on year, pretty much indefinitely into the future.

But scientists are warning that these hopes for nature-based solutions may be confounded precisely because the planet is already warming so fast. At Climate Week in New York in September 2023, Johan Rockström, Director of the Potsdam Institute for Climate Impact Research, warned politicians that 'we are already seeing massive cracks in the resilience of the Earth's systems, both on land and in the oceans'. What the Institute's report showed was that the amount of CO_2 absorbed by terrestrial systems, e.g., forests, soils, peat bogs, in 2023

3: CLIMATE SCIENCE

was close to zero. Severe droughts and massive forest fires had contributed significantly to that failure.

You don't have to be a twenty-five-year-old political prisoner to recognise that dependence on magical thinking of this kind has a lot more to do with powerful strains of 'hopium' than any kind of serious science.

4. SURELY IT'S NOT THAT BAD?

Mrs Thatcher was famed for telling her Ministers to bring her the good news rather than constantly bombard her with the problems. But when it comes to the climate crisis, there isn't much good news. And most politicians still feel deeply uncomfortable in being obliged to confront all the bad news.

Unfortunately, climate science doesn't really work like that. Empirical reality has literally nothing to do with the optimism or pessimism bias of individuals. Science observes, measures, records and communicates about the way in which the laws of physics are transforming every square inch of our planet – on land and at sea. You can't negotiate with those laws just because the wholesale transformation going on is likely to leave you feeling a bit depressed.

It was intriguing to see how politicians reacted to some of the extreme weather events in 2023 and 2024. In October 2024, for instance, the World Health Organization reported that around 50% of the Earth's landmass had been affected by at least one month's worth of extreme drought, with devastating impacts on human health. Heat-related deaths were up by 167%; sand and dust storms increased by a third; 150 million additional people ended up in 'food insecurity', often not knowing where their next meal would come from. Yet there was no political traction whatsoever.

Whilst admitting that extremes of this kind are disturbing, politicians often point to the fact that 2023/2024 was an El Niño year so perhaps things are not quite as bad as they may appear. The oscillation

'We really are facing an apocalyptic event. We need people to be scared about this – to be scared shitless. But if people feel it's already too late, they're not going to do anything about it. So we have to demonstrate that it's *not* too late.

'I'm still optimistic. We're such an intelligent species – although you often wouldn't know it! I try and avoid doomscrolling. For me, it's about balance. People are always going to be sad about something, then happy about something; so while I want people to be scared shitless, I also want them to stay positive about all the good things in life.'

Harrison Donnelly

'In my first year at uni, Extinction Rebellion seemed to cave in to public pressure, and decided it wasn't going to do any more direct action. I was studying some directly relevant modules and became utterly depressed about my future – it was hard to feel motivation, having come to Oxford to study biology and discovering that all the advice from scientists was being ignored. There seemed to be nothing opposing what the politicians were doing – or weren't doing – with species disappearing all the time. I was feeling isolated at that time; it seemed that nobody really cared about things, and they weren't even willing to make minor changes – to their diet or the way they lived.

'There are a lot of us in Just Stop Oil who have to find the best way of balancing activism with their degree. I want to stay in academia and, for me, the two things are not completely separated. I now think about research priorities in a different way, thinking about ways in which one can make a positive contribution – my work on how beaver re-introductions impact ecosystems, for example, to reinforce the case for rewilding throughout the UK.'

Ollie Sworder

between El Niños, warmer phases, and La Niñas, cooler phases, based on sea surface temperatures in the Pacific Ocean, is complex science. But somehow whichever phase we're in seems to provide a lot of wiggle room for the it's-not-that-bad merchants!

El Niño was also blamed by some for the crazy temperature 'anomalies' recorded both on land and in ocean surface temperatures. Scientists are still arguing – true to the scientific method! – about whether this was a 'blip', 'blippish' or 'evidence of a much more profound systemic shift'. We non-scientific mortals go there at our peril, but most politicians will unerringly go to the best possible interpretation of what's happening today. Don't let go of that comfort blanket!

'But what gives them the right to do that?', Just Stop Oil activists ask. 'Who gives a shit if it's all down to El Niño or nothing to do with El Niño at all? There will be another El Niño along in a year or two, during which time emissions of greenhouse gases will have continued to rise and concentrations in the atmosphere will have continued to build, intensifying the same old cycle down here on Planet Earth.'

Indeed. It's only 'not that bad' if you look at it with blinkers on, and over the shortest possible time cycle.

Behind all these statistics are the stories of individual people. In October 2024, Valencia in Spain received a year's rainfall in just eight hours. This caused devastating flash floods, turning streets into rivers, sweeping people, many of them elderly, to their deaths – almost half of all victims were aged over seventy. At least 224 people lost their lives. They died in underground garages, in their cars and in care homes. Cars were 'thrown around like toys'. Many homes were destroyed and many more rendered temporarily uninhabitable. The clean-up took days to get underway and the smell was unbearable. Rescue workers cried as they painted purple crosses on cars to indicate where someone had died.

Every single climate disaster comes down to personal tragedies of this kind, to immediate trauma, to recurring nightmares, to prolonged suffering.

'I'm not sure there's going to be one particular moment which brings people together, like Rosa Parks refusing to give up her seat on a bus during the civil-rights struggle. People seem to be getting used to ever more extreme disruption – floods in Pakistan, wildfires in Australia, heatwaves everywhere. It's difficult to believe that so many people now think these things are 'normal'!

'I suspect it's going to be something like a massive crop failure that really shifts the dial. So few countries produce so much of the wheat on which we depend, and it's not difficult to imagine a climate shock which would cause harvests to fail in two or three countries at the same time – a so-called "multi-breadbasket failure".'

Alex

'As part of my degree, I had an internship at WRAP, the Government's advisory body on waste. And I began to see some of the contradictions at work in an organisation like this – making tiny little improvements against the backdrop of a full-on climate crisis. Other staff in WRAP shared my high-level climate anxiety, but I was shocked to hear them say things like: "I feel I have to leave all that stuff at home to get my work done" – in an organisation that is supposed to be facing those challenges head on! Why are we talking about marginal improvements in recycling rates without any real understanding of the deeper situation? And I guess that was where my game plan at that time (finish my PhD, find myself a good job in a well-respected environmental organisation, and change the world in that way) literally died. I just couldn't pretend any longer.'

Sean

3: CLIMATE SCIENCE

5. THE TRAGEDY OF THE HORIZON

Ten years ago, in September 2015, Mark Carney, then Governor of the Bank of England, gave this warning:

> Climate change is the tragedy of the horizon. We don't need an army of actuaries to tell us that the catastrophic impacts of climate change will be felt beyond the traditional horizons of most actors – imposing a cost on future generations that the current generation has no direct incentive to fix. Once climate change becomes a defining issue for financial stability, it may already be too late.

We are all bit-part players in this tragedy, not just the politicians. But some have much greater agency than others. And with that greater agency comes greater responsibility. And here's the thing that those with greater responsibility and agency need to realise: what was once seen as a potential crisis out there on the distant horizon, decades away, has demonstrated a nasty tendency to end up in our own back yard much sooner than anticipated.

I've followed news about climate impacts for longer than the lifespans of the activists I've been working with on this book. Throughout that time, there has been a disturbing leitmotif in the interpretation of climate science: whatever predictions scientists may have made in the past about future impacts – extreme weather events, temperature increases, rising sea levels, disappearing ice and permafrost – the projected time span within which those changes are predicted to occur has steadily reduced over the years. In forty years time, for example, has become in twenty years time; predictions have become medium term rather than long term. I can't recall a single occasion when I've been reassured by new evidence telling us that some projected change is now much further away than we had once thought.

That is certainly the case with the threat of rising sea levels. The consequences of even a 1-metre average rise in sea levels is already beyond our imagination. Two thirds of the world's largest cities are located within a metre of sea level. And scientists have warned that if we exceed the 2°C average temperature increase which is now seen by many as all but inevitable, then we could see as much as a 3-metre rise in sea levels by the end of the century.

Even five years ago, most climate scientists would have been very cautious in their estimates of future sea-level rises, particularly those involved in the work of the IPCC. That caution is now disappearing as fast as the melting ice caps that contribute to rising sea levels. Contemplating a decade of failure in reducing emissions since the 2015 Paris Agreement, it is now commonplace for scientists to acknowledge that temperatures will rise by between 2.5°C and 3°C by the end of the century. And some, including the redoubtable Professor Jim Hansen, believe that the 2°C threshold is already 'deader than a doornail', just as 1.5°C is, with a temperature increase in excess of 3°C looking more and more likely.

For most scientists, this is indeed uncharted territory, and most find it really hard processing the emotional implications of such scenarios. They're also startlingly reluctant to acknowledge just how badly wrong mainstream science has been in its earlier projections, and in its dogged dependence on integrated models that now look completely broken.

One way or another, we're all still playing our parts in this tragedy of the horizon, precisely because what happens out there is indeed beyond our imagination. But that admittedly arbitrary date of 2100 is inevitably much more present in the imagination of a twenty-five-year-old than it is in those who know they'll be dead by 2050 – i.e., most politicians today. That means that short-term crises continue to dominate the political space.

Our political systems are simply not set up to address such consequential scenarios. But Just Stop Oil activists would argue that serving politicians have a responsibility, at the very least, to engage seriously enough with this potentially catastrophic outcome – rather than acting as if there was literally 'nothing to see here'.

CRESSIE GETHIN

I'm twenty-two and grew up in Herefordshire in a family of musicians. I took a gap year, mainly at home due to the Covid pandemic, which is when I started to become more politically engaged and more aware of the climate crisis. I got involved in Just Stop Oil in my first year at uni but dropped out in the second year to focus on JSO full-time.

INVOLVEMENT
April 2022: Actions at oil refineries.
July 2022: Gantry banner drop, blocking M25, in response to 40°C heatwave.
November 2022: Similar action – this is the action for which I was in prison, until released on Home Detention Curfew in April 2025.
June to November 2023: Slow marching in London. Lots of arrests which never went anywhere; remanded in prison for three weeks in November.
2024: Stepped back in 2024, with two Crown Court trials to deal with, to decide how best to take forward this work.

MOTIVATION
I believe this is the right thing to do. The world is profoundly out of balance, with the material comfort of some resting on the violent oppression of others and all non-human life. To not resist this makes me sick; I cannot separate my own well-being from the well-being of others, and not confronting the injustice is as much a violation of myself as of everyone else. I believe that nonviolent action is the most effective way of challenging structures of domination without inadvertently reproducing them.

INSPIRATION
My dear friend Jon, who I met when I was very new to activism. He is one of the bravest people I know, and I have the utmost respect for him. One thing he has taught me is to 'widen my zone of consideration'. He trusts me to hold him accountable in the same way as he does for me, always listening when I'm struggling, without trying to solve problems for me.

IN NATURE
There's a woodland near where I grew up which at one point opens into a glade where the ground is covered in moss, where there is an extremely large, old oak tree. I used to sit in its branches when I was young. It's not a 'pretty' tree – it's gnarled and covered in scars, with some of its branches damaged by lightning. But it's beautiful and powerful. There's also a bunker nearby which fills with the leaves of the tree in autumn – perfect for diving into! But I dive more carefully these days to avoid disturbing any hibernating hedgehogs!

QUOTATION
'We have not made a single gain in civil rights without determined legal and nonviolent pressure.' Martin Luther King

For example, dehumanising people by thinking of them as 'enemies', can be unhelpfully corrected by swinging to the other extreme, and failing to call out oppressive power hierarchies for fear of dehumanising those in power. This quote illustrates the third way between these two.

RESOURCES
Healing Resistance by Kazu Haga. This book was what helped me move beyond thinking instrumentally about nonviolence as a tactic, as a tool, and to see it as a way of life, a unifying narrative. It's allowed me to move beyond a fixation on the immediate outcomes of my action, and to understand the power of doing what's right even when there is no obvious cause and effect. I believe that this has had a positive influence on the choices I've made, putting them more in alignment with justice. I found that thinking this way is also profoundly healing for me, and whenever I see or feel this depth on

nonviolence, I feel a warmth inside that is incredibly nourishing, and a desire to share it with others.

WHAT LIES AHEAD?
I hope that more and more people will start practising nonviolence, challenging power, strengthening communities on a large scale. I also hope that more and more international links will be forged between people engaged in the struggle all around the world – the more we can do this, the better our chances of resisting the pull towards fascism as conditions become more and more unstable.

What I hope for is that social and ecological collapse does not end in all-out war, and that there will be more people who protect each other rather than killing each other. I have faith that true deep nonviolence that is at once challenging and loving will always eventually lead to something good.

DANIEL HALL

I was born and brought up in Croydon. I learned about climate change in primary school, but just assumed everything would be 'under control', so I parked it, had a child at twenty-one, joined the family business, got married and we now have two children.

In some ways, I grew up as an 'angry young man', part of that whole 'left behind' story, raging at the injustice of 'the system', the corrupted political elite. As a millennial, it was the

Iraq War we went through when I was at school, the 2008 financial crash as I was leaving school, and then austerity throughout my adult life. So when Osborne and Cameron came out as 'remain voters', I gladly voted 'leave', hoping it would destabilise the Government and give that elite a shake.

INVOLVEMENT

With two very young children, XR really got both of us thinking about their future. My partner got involved in supporting XR early on, but for a while I think I was still stuck in some sort of 'soft denial'. Then I connected very deeply with the first Insulate Britain protest and began to feel hope that something could be done about this.

After that, I really got into Just Stop Oil, with four arrests for slow marching, in Parliament Square, for which charges were later dropped, and Cromwell Road, for which I stood trial under Section 7 of the 2023 Public Order Act, together with Phoebe Plummer and Chiara Sarti. We were found guilty, but I was given a community rather than a custodial sentence, with a Criminal Banning Order that restricts involvement in further actions.

I still have one Section 7 prosecution outstanding, with a Crown Court trial now deferred until 2027 – that really is 2027!

INSPIRATION

I've been deeply influenced both by the Freedom Riders and the Suffragettes. These days, one has to interpret what the Suffragettes did within the context of that historical moment – including some of their violent tactics. But theirs was the true revolutionary spirit, going out knowing what is morally right, knowing you're going to be hated, knowing you're going to be hurt. That is what it means to be human.

I've also been reading a lot about Rojava (see p. 147) and the extraordinary determination of the Kurdish people in that part of Syria to establish a genuinely fair and equal society.

MOTIVATION

Nobody with young children needs any further reason for getting involved.

And nobody who listens to what the scientists are telling us needs any further reason for getting involved urgently.

IN NATURE
The nature connection for me was never as strong as for some JSO colleagues, but I love being here on this little patch of land, surrounded by trees, and living as simply and authentically as we can.

QUOTATION
These lines attributed to Lenin always remind me how important it is to stay focused:
> I know of nothing better than the Appassionata ... But I can't listen to music very often ... I want to say sweet, silly things, and pat the little heads of people who, living in a filthy hell, can create such beauty. If I keep listening to it, I won't finish the revolution.

RESOURCES
I've already mentioned the Freedom Riders and can warmly recommend a short film on YouTube called AMERICAN EXPERIENCE: Freedom Riders: The Tactic. That whole period in history, with people stepping up so selflessly and courageously, provides such a powerful antidote to the kind of anger that a lot of people take refuge in. It shows so clearly why confrontational but entirely non-violent direct action really works.

WHAT LIES AHEAD?
There won't be just one moment where everything gets magically sorted! I think it will be more like a rolling revolution – two steps forward, one step back, with a lot of inevitable backlash already going on today, even with things like trying to persuade people to eat less meat.

So, we just need to adopt a revolutionary mindset, to be ready for everything – anything is possible as long as we stay on it. And just keep on discovering what works!

'One thing that makes me very sad is the depoliticisation of universities and students. It's been such a calculated process ever since Margaret Thatcher. That often comes to mind when I'm asking myself how I can be most effective in my civil resistance. A lot of people who know a bit about the climate crisis, don't want to talk to me about those things – out of fear, perhaps, of having to recognise the real situation. That can be very isolating. But I'm not here to judge anyone in terms of where they themselves are.'

Paul

'This is a huge political struggle, but the way we respond is always personal. When I first decided to get involved, my younger sister was pregnant in Italy, which was experiencing record-breaking heatwaves, that caused her significant health problems. I really connected to that.

'I've never been a conventional environmentalist. For me, it's much more about human rights, government corruption and how we have so little respect for some people's lives. I was basically an annoying liberal, thinking I was much more radical than I really was, just chirping away about things without ever following through. I came to uni in London in 2018, and Extinction Rebellion's first bridge occupation was suddenly all around me, very good-natured, celebratory even. I reacted straight away – 'better go vegan', that sort of thing, banging on about a billion people being displaced because of extreme climate chaos – but it was all a bit superficial. It got a lot more serious later on.'

Chiara

4: CLIMATE POLITICS

'Inequality demands oppression. The more concentrated wealth and power become, the more those who challenge the rich and powerful must be hounded and crushed.'

George Monbiot

CLIMATE CHANGE is not, and never has been, 'an environmental issue'. It is not, and never has been, 'an issue' of any kind. It is an unfolding physical and geopolitical reality that is already affecting the lives of the vast majority of human beings, and will, in the not too distant future, become the single most significant influence determining the future of our entire species.

You wouldn't know it! Be it through deliberate intent, as with political parties or mainstream media conglomerates owned or controlled by plutocrats whose interests are entirely aligned with today's fossil-fuel companies, or by an extraordinary combination of ignorance, inertia, indifference or a straightforward 'lack of agency' on the part of the majority of citizens in nations rich and poor, *a true understanding of the threats posed by accelerating climate change is restricted to relatively tiny minorities of people in every nation.*

Never have so many been so ill-informed about matters of such material significance to them personally, to their communities and countries, and to the whole of humankind.

Yet climate change, or global warming, as it was originally described, has been in the news for decades. In the run-up to the Earth Summit in Rio de Janeiro in 1992, where the UN Framework Convention on Climate Change was brought into being, via the big climate conferences held at the end of each year, through to the predictable announcement that 2024 had been the hottest year ever, climate change has always been there or thereabouts on nation states' 'to-do' lists and in news editors' in-trays. However, genuine cut-through to the general public has remained minimal, with limited political impact. Indeed, the fact that climate change has been part of the public discourse for so long has left many people thinking that it must somehow have been 'dealt with'.

Pretty much everything that Just Stop Oil activists feel called on to do today is a direct consequence of commitments not honoured, decisions not taken, and well-meaning processes having been abused throughout that time. The full impact of those three 'wasted decades' is only now becoming apparent. It really is no exaggeration to say that had we moved forward, steadily but purposefully, on all the different elements captured in that ground-breaking Framework Convention in 1992, not only would we not be in the emergency we are in now, we would still have a very reasonable prospect of avoiding what have gradually become the unavoidable consequences of so little progress having been made since then.

Politicians have a very bad habit of listening only to those aspects of climate science that can be accommodated within today's dominant worldview with its fixation on market forces and constant economic growth. It matters little that climate scientists have been pointing out to them for decades that the laws of thermodynamics will always trump the so-called 'laws' of the market – it's just a question of *when* that will happen, not *if*.

Now, in 2025, I think we can safely say that the 'when' question has been partly answered: a great deal earlier than most customarily conservative scientists have been suggesting. As summarised in the previous chapter, the pace of change, as measured in real time, month-on-month data sets gathered in from critical ecosystems all around the world, has been astonishing. Scientists have been struggling to communicate just how astonishing the pace of change has been, both to the politicians and the general public.

Most of the Just Stop Oil activists I've interviewed for this book are in their early to mid-twenties. They look back over the last three decades as so many wasted years, going on throughout their short lives.

4: CLIMATE POLITICS

It's important for people to understand why we're in such a crisis today, and why the last six or seven years of climate campaigning should be seen as radically different to everything that went before. It helps to see things in terms of overlapping timelines involving four key constituencies of interest:

- the climate scientists, who've been there from the start;
- the politicians, who are only now beginning to get their heads around what's really at stake;
- the many different representatives of civil society, non-governmental organisations (NGOs), community groups and so on;
- and, of course, the fossil-fuel industry itself and their 'enablers' – the banks, insurance companies, asset managers and legions of right-wing media.

I'll conclude the chapter by looking at what this all meant for the vexed relationship between Just Stop Oil and the Labour Government.

1. THE SCIENTISTS

Professor Jim Hansen's testimony to the US Congress in 1988 is usually recognised as the moment when the somewhat esoteric science about atmospheric chemistry crossed over into the public domain. Thirty-seven years on, Hansen is still just as much at the heart of today's scientific controversies. He's also out there as an unapologetic climate activist – he's been arrested on a number of occasions – and can be pretty withering about the majority of his colleagues remaining so calm in the middle of a rhetorical war of words! For the purposes of this book, there's no point going back further than Jim Hansen's original testimony in 1988, although it's sobering to realise that the basic physics and chemistry involved in climate change were well understood by scientists as far back as the 19th century.

'One of the main demands of This is Rigged is for community food hubs, to nationalise this part of the food system. Scotland has signed up to a European law that says everyone has the right to food – three meals a day. The problem with our Government is that it doesn't walk the talk. A lot of MSPs (Members of the Scottish Parliament) will say they agree with this, hypothetically, but they won't put any work into making it happen.

'We have to start asking how we're actually going to feed people in the future. Think how unstable our food and water systems are going to be, as food prices continue to rise, and only those who can afford it will be able to eat. So we need a food system that is built for people – one in four people in Scotland already lives in food insecurity, and that's before we take into account the impact of wildfires, droughts and floods. What's it going to be like in ten years? In twenty years? Fifty years?

'I used to be really into Scottish nationalism and independence. Now, it's not that I don't want Scottish independence, but what if we just end up with another government that has no real care for its people and the planet, no proper democracy, just swapping bad for slightly less bad. I've lost faith in "democracy".

'What we need is much more like community power, with people having power to make decisions in the places where they live. There's far too much that is top-down. Give people proper decision-making powers, and they generally make good decisions, because they know the places best. We need to completely rethink how decisions are made, and what level democracy works best at. It's about devolving power and putting trust in people again to make the right decisions. Then I think they will surprise you!'

Eilidh McFadden

4: CLIMATE POLITICS

The science of climate change isn't really about individual scientists; it's about the scientific method already described: 'The gathering, processing and promulgation of evidence to advance overall understanding.' That is what led politicians to set up the Intergovernmental Panel on Climate Change, to sit a little above the fray and to provide a mechanism for crunching increasing numbers of peer-reviewed papers covering every conceivable facet of climate science.

It has the additional role of brokering that science with the politicians, through its regular but unbelievably laborious 'Assessment Report' process. Not a single one of those reports has ever emerged unscathed from the IPCC; all have had to secure the approval of all UN member states – including countries such as Russia, Saudi Arabia and the USA. For me, that 'remorseless politicisation of climate science' has largely undermined the IPCC's credibility. Its report in 2018, 'Global Warming of 1.5°C', came as close as it ever has to setting out just how bad things have become, but still couldn't quite get the words out, and left plenty of wiggle room for the politicians to evade their obligations.

The IPCC's current Chair, Professor Sir Jim Skea, has opted to be particularly diplomatic in the way he seeks to balance this science/politics dilemma. Unfortunately, he's also chosen to attack the reputation of the climate movement's Radical Flank. In August 2023, he accused Just Stop Oil campaigners of 'paralysing' the fight against climate change: 'if you constantly communicate the message that we are all doomed to extinction,' he stated, 'then that paralyses people and prevents them from taking the necessary steps to get a grip on climate change.' This is such a fatuous misrepresentation of what Just Stop Oil actually said during its short-lived existence. While Professor Skea himself now acknowledges that it is 'almost certainly too late' to restrict the average global temperature rise to below 1.5°C, he's keen to reassure people that this doesn't necessarily constitute an existential threat to the future of humankind. At the same time, he's admitted to being a bit worried that 'we are currently on track to a 3°C temperature increase by 2100', which I think pretty much everyone would recognise does indeed represent an existential threat!

It's not surprising then that Just Stop Oil activists relied on more independent scientific experts. For them, the bottom line here was pithily summarised by Professor Hans Joachim Schellnhuber: 'Political reality must be grounded in physical reality, or it's completely useless.' Indirectly, that would seem to imply that the highly politicised conclusions of the IPCC may also be 'completely useless'. They certainly run the risk of paralysing politicians the world over with a false sense of reassurance.

2. THE POLITICIANS

A word in defence of the politicians before trying to explain why Just Stop Oil campaigners are so embittered about their collective failure: dealing with the Climate Emergency is not easy. There is no manual, no playbook, no precedent. The future of the entire global economy is at stake – as is the very future of humankind.

I still hear laboured comparisons between the excellent job that politicians did in the 1980s in sorting out the potentially existential threat caused by the thinning of the ozone layer, and the total mess that they're making sorting out the potentially existential threat of accelerating climate change. Unfortunately, these two potentially existential threats are not remotely comparable.

It's true that if politicians back then had ignored the irrefutable scientific evidence of ozone depletion, caused primarily by the pervasive use of chemicals like chlorofluorocarbons (CFCs) in refrigeration, aerosols and so on, and if they hadn't realised they could negotiate with the manufacturers of those chemicals (ICI, DuPont, Atochem and others), given that perfectly good alternatives to those CFCs were available, and if they hadn't risen to this massive challenge by drafting, negotiating and signing off a binding international agreement, The Montreal Protocol, in a very short period of time, then life on Earth today would be very different – and very unpleasant – without a properly functioning ozone layer.

4: CLIMATE POLITICS

But climate change isn't like that! The use of fossil fuels permeates every single aspect of our lives today. Our material standard of living currently depends on the availability of those fossil fuels and – let's be honest about this – the lives of billions of people have been significantly enhanced by our access to relatively cheap, extremely efficient energy sources of this kind. Trillions of dollars have been invested in a complex infrastructure that continues to make these benefits available, and trillions of dollars in economic value will be foregone, i.e., will no longer contribute to the kind of economic growth on which we are still so dependent, when we get out of fossil fuels. But that is what we now have to do. And as quickly as is humanly possible.

And that is because our continuing dependence on fossil fuels will, in time, bring an end to human civilisation. No ifs or buts – that's what the science tells us. And no politicians of any generation, including those who went through the two World Wars and the Cold War of the twentieth century, have ever had to address a rapidly evolving emergency as challenging and all-encompassing as this one.

All that is worth bearing in mind – but that's as far as my political empathy goes! The harsh truth is that politicians are making a catastrophically bad job of addressing this challenge – through scientific illiteracy (most of them know literally nothing about the true nature of the Climate Emergency, and are remarkably reluctant to overcome that deficit); through inertia ('We really have got enough to be dealing with already'); and through corruption of varying degrees, all the way from accepting donations from fossil-fuel interests to being content to be 'bought' – literally – by fossil-fuel companies, as is still the case with the majority of US politicians today.

When I say that fossil fuels permeate every single aspect of our lives, that's what it feels like politically – like an ongoing, permanently repeating 'oil spill' seeping into every nook and cranny of our political ecosystems. The clean-up process is going to have to be radical, comprehensive and inevitably very painful.

The whole sorry saga of two generations of politicians having been

'Maybe it's how my neurodivergent brain works, but whenever I'm studying social movements from the past, I imagine myself in those situations. When I was sixteen, I participated in a Holocaust Remembrance programme, visiting Auschwitz and other concentration camps. Once you've seen that, how can you stay silent on present-day genocides? And if I hadn't been involved in activism when I was studying the Suffragettes, I would have gone crazy thinking about all those incredible women who put everything on the line with such extraordinary courage. I've always found it easy to empathise with people in these situations, to connect emotionally.'

Emma de Saram

'What still annoys me today is that so little seems to be changing – even now. The people who really make me angry are those who endlessly tell everybody how much they care about the environmental crisis, and then don't actually *do* anything! They just end up criticising Just Stop Oil for taking the actions that we do. I can sort of understand this, given how terrifying it's become to go up against the Government, but we need far more people stepping up and really trying to get their heads around what the climate crisis is going to mean in the future.

'But Just Stop Oil is not the "be all and end all" of activism today. It's not going to be around forever, and it might evolve into something else quite soon. That's a good thing. I feel proud of what Just Stop Oil has achieved, especially given the lack of progress before ER and JSO, but it's obviously had to adapt to very different circumstances after the Tories introduced their new anti-protest legislation.'

Cole

'missing in action' is now being played out through the arcane proceedings of the UN's Framework Convention on Climate Change, its permanent Secretariat, and an annual jamboree known as the Conference of the Parties, i.e., the signatories to the original convention. This means nothing to the vast majority of people who have had literally zero connection with this process over the past thirty years, but every now and then there is a 'big COP' that succeeds in crossing some kind of visibility threshold, and has an impact on at least a few million people around the world.

Just Stop Oil activists were unanimous in their condemnation of the entire 'COP charade'. In 2021, a number of them attended COP26 in Glasgow, which did nothing to allay their scepticism. Astonishingly, this was the first COP where there was any kind of focus on fossil fuels, with a massive row about the difference between 'phasing out' and 'phasing down' the use of coal power, even though there was still no explicit reference to the use of fossil fuels being the principal driver of accelerating climate change – nearly thirty years after the Framework Convention had come into being.

The gap between political reality and physical reality could not have been clearer, to use Schellnhuber's terms. On the one hand, the 'stubborn optimists' at COP26 kept banging on about it not being too late to restrict an average temperature increase to below 1.5°C – as explained in the previous chapter; on the other hand, more and more 'climate realists' were pointing out even then that the 1.5°C threshold had almost certainly already been breached – as became even clearer at COP28 in the United Arab Emirates and COP29 in Azerbaijan, two totally corrupt petrostates with absolutely no intention of slowing their production of fossil fuels.

There's no residual mystery about this COP charade. The proceedings of the Framework Convention are dominated by coal-, oil- and gas-producing nations, and by coal-, oil- and gas-consuming nations. Their national interests are inextricably intertwined with the interests of those industries. For dictatorial petrostates like Russia, Saudi

Arabia, the UAE, Qatar, Iraq, Azerbaijan and Venezuela, there is no contradiction at work here: hydrocarbons *are* the state. But for most Western democracies, the whole COP process has been nothing more than a thirty-year exercise in 'systemic cognitive dissonance', with politicians ardently declaring their determination to reduce emissions of greenhouse gases just as fast as possible whilst continuing to serve the interests of what is by far the world's most powerful industry: fossil-fuel production.

Nowhere is this clearer than in the USA, where climate activists confront a multi-trillion-dollar industry that permeates every single aspect of American politics, economics and culture. No US President since Jimmy Carter, who at least understood what he was up against, has even attempted to begin to unravel this sick symbiosis, and that includes both Barack Obama and Joe Biden, both of whom learned how to 'speak fluent climate' without in any way challenging the status quo.

It's not a totally failing scene. The European Union has consistently provided real leadership at these conferences throughout this time – although that too is now under attack with increasingly powerful right-wing parties (including the Alternative für Deutschland, which came second with an astonishing 20 per cent of the vote in the German election in February 2025), pushing back against ambitious targets both in the EU parliament and in many individual member countries. Beyond the EU, year after year, the Association of Small Island States has done everything it could to keep the process 'honest'. But with that symbolic 1.5°C threshold now acknowledged to have already been breached in 2024, it's just a question of how long they will survive as viable nations – no longer if they'll survive. They won't.

More and more people now reluctantly acknowledge that this is an inherently dysfunctional process that continues to fail year after year. The rationale for sticking with it – 'If something like this didn't exist, we'd have to invent it' – still holds sway with most politicians and NGOs. But the years just slide by, with all the indicators continuing to head in the wrong direction.

OLIVER CLEGG

When you're in police custody, there isn't much to do but think about why you're in police custody. You can read or sleep – my preferred activities – but you'll inevitably end up thinking, 'Why did I end up here?' If you're lucky, the answer might be: 'I'm here because I've just been disrupting the fossil-fuel industry, the single most destructive industry in the world.' But after one memorably shite protest, the thought went differently. I was arrested after half an hour blocking the road out of an oil terminal. I couldn't sincerely believe that I'd crippled the oil industry with half an hour of marginal disruption. I also couldn't believe that I had an absolute moral obligation to go on disrupting the oil industry. What would that mean, if I was under an absolute moral obligation? Would I be required to blow up the Shell HQ? Would I be obliged to go on hunger strike until all oil drilling had stopped, and end up dying as some sort of climate martyr? I'm not cut out for martyrdom!

I did settle on a reason why I had got myself arrested though: activists of the past did braver things. The gay activists of the 1970s could have remained closeted. It wouldn't have been nice, 'living a lie' and all that, but it would have been nicer than being openly gay at that time. And if those activists had stayed in the closet rather than organising protests, my life would be measurably worse. Think of a group like the Gay Liberation Front, which, as well as fighting homophobia, fought racism and capitalism. It's hard to imagine that its activists intended to push for gay equality so that a gay man born in 2003 could sit around having mediocre sex whilst remaining politically unengaged. I owe the activists who made my life better a continuation of their fight for a better world.

I've been quite aware of my own lack of gratitude recently. I have a nice life, I really do. I was in a Waterstones café a while ago, and was annoyed with myself that all I was doing was killing time. The big, momentous action I was planning was in the future, and its fate was really in someone else's hands. So all I could do was sit there killing time, reading my book, drinking my coffee and eating my cake. Not feeling good about things. And then I thought, 'How fucking pompous! How dare you sit here feeling all sorry for yourself with your bourgeois life? In a café, inside a fucking bookshop! Where's your gratitude?'

Again, I urge you to try to see that chronic and continuing diplomatic failure through the eyes of young Just Stop Oil activists. Against such a backdrop, the idea that they're the ones responsible for 'paralysing' the general public in responding appropriately to the Climate Emergency, as suggested by establishment figures like Sir Jim Skea, Chair of the IPCC, becomes more and more preposterous.

3. CIVIL SOCIETY

By and large, over the three wasted decades, the overall positioning of NGOs, progressive think-tanks and community organisations, has stayed in line with the official scientific consensus – and proved itself to be consistently incapable of making any real impact on what is happening politically.

Extinction Rebellion didn't just emerge out of a vacuum in 2018. It had been clear for a long time that NGOs were 'playing down' the seriousness of what was emerging on the front line of climate science, largely as a consequence of sticking with the laborious and heavily politicised process of the IPCC. Tellingly, XR's first demand was 'Tell the Truth', a message directed just as much at the mainstream NGOs as at the politicians.

That original call to 'Tell the Truth' from XR was powerfully amplified six years later in 2024 at what became known as the trial of 'The Whole Truth Five' – Daniel Shaw, Louise Lancaster, Lucia Whittaker De Abreu, Roger Hallam and Cressie Gethin – at Southwark Crown Court. Judge Hehir had ruled that the defendants' motivation in planning an action to disrupt traffic on the M25 was of no relevance to the jury, and that their 'opinions' about climate change were inadmissible and should not be heard in court. As discussed on page 39, Hallam was sentenced to five years; the others to four years.

But the truth, and today's truth-tellers, can no longer be silenced. In a world where the extreme effects of climate change are already being experienced on the ground, week after week, in country after country, extrapolating from what we see and feel today to where we'll find ourselves in the not too distant future, with a minimum of a further 1°C temperature increase already baked in, can no longer be dismissed as 'sensationalist' or 'scaremongering'.

From a campaigning point of view, this should make it easier to ramp up the pressure on both politicians and businesses. Both outright climate deniers and all the equivocating 'delayers', described by Michael Mann as the 'new deniers', have seen much of their credibility shredded over the last couple of years. Where once organisations like the Global Warming Policy Foundation or the Institute of Economic Affairs might have been able to make use of a wafer-thin veil of pseudo-science to trot out alongside their lies, that has now largely gone. They stand revealed as mendacious, self-interested ideologues, vainly attempting to argue that the so-called 'laws of the market' should still, despite all the evidence, take precedence over the laws of physics.

Unfortunately, however, these are still influential people, with powerful financial backers, capable of causing a great deal of trouble in the future. And with Donald Trump back in the White House, and right-wing, populist parties gaining more and more ground in Europe, earlier hopes that climate denialism was dead and buried were clearly premature.

This is why representatives of civil society have all got to up their game. Unfortunately, some are dangerously close to becoming complicit in the continuing destruction of the natural world and climate breakdown (as I explain in the next chapter), dependent as they are on donations from the super-rich, and on maintaining their credibility with establishment business, political and media leaders.

And for all those in the Green Movement who remain critical of the Radical Flank, still represented by organisations such as XR, Fossil Free London, and Green New Deal Rising and This is Rigged – and formerly by Just Stop Oil – criticising their tactics on the grounds that the general public is being alienated, with the media being given huge amounts of

'Being in prison has gifted me time, headspace and perspective to reflect on things – first, when I was on remand for slow marching and now, serving a sentence for conspiracy to cause a public nuisance. It's allowed me to think more deeply about which strategic considerations to prioritise, where JSO has got things right, and where it hasn't. When it comes to leadership, I'm inspired by the vision of "leaderful" movements. I have had to step up myself, giving public talks, being proactive in group settings, even though that doesn't come naturally to me. Particularly as a young person, I don't like to claim expertise that I don't have. What I can do is to speak from my own lived experiences, and to centre the voices of otherwise silenced people – who are usually the ones with the wisest things to say.'

Cressie

'Electing a Labour Government has persuaded many people that things are now going to get sorted, in a way that was never going to happen under the Tories, but there are so many ways in which politicians provide "false reassurance". It was easy for Parliament to declare a Climate Emergency, but then nothing really changed. People think "Oh, that's good, now things will really start changing", which just serves as a pretext not to get stuck in more actively. Even though many of them never had any intention of getting properly stuck in anyway!

'Deep down, there's not much difference between the economic policies of Labour and the Tories. It's the same old neoliberalism, just dressed up differently. I know a lot of people don't see it like that, but one advantage of our kind of climate activism – and possibly because I'm somewhere on the spectrum! – is that we have to see things in black and white, not focus on all those grey areas that provide excuses for not doing what needs to be done.'

Daniel H.

ammunition to undermine the credibility of the Moderate Flank, I must respectfully suggest that you're now even more off the pace than you were before, never truly ready to confront the 'whole truth', and still in danger of finding yourself on the wrong side of history.

And that may be because you're still standing too close to the real villains of the piece.

4. THE FOSSIL-FUEL INDUSTRY

I've devoted the next chapter to what I call the fossil-fuel incumbency, so I can be brief here, simply highlighting the hugely destructive influence that this industry has had on civil society.

I know a bit about this, and have written before about the relationships that Forum for the Future had with both BP and Shell over many years. When we worked with them (especially with John Browne as CEO at BP) both companies had declared a readiness to transition away from being 'pure-play oil and gas companies' into 'integrated energy companies', with far more renewables and bioenergy in their portfolios – in effect, bigger versions of the erstwhile Danish Oil and Gas Company (DONG) which very successfully transitioned completely out of oil and gas and into renewables (particularly offshore wind) as the company we now know as Ørsted.

But everything we did with them – masterclasses for senior executives in the UK, USA and the Netherlands, climate modelling over different time periods, one-to-ones with the 'bosses', challenge sessions with new recruits etc. – was, in the end, ignored. With the benefit of hindsight, of course it was. Was Forum for the Future 'corrupted' by these working relationships? I would argue categorically not. They paid us for the work we did, as a charity, just as they worked with and paid many NGOs around the world. But were we naïve in persisting for so long in the illusion that we could somehow have an influence on such massive, profit-maximising companies? We absolutely were.

'I don't use the word lightly, but so many decisions being made today are evil – about the climate crisis that will cause untold suffering to billions of people in the future, and about the unfolding genocide in Palestine. Even the impact of poverty and austerity on the lives of millions of people in the UK is evil, the result of decisions made by a tiny number of very privileged people.

'In prison I read *The Shock Doctrine*, by Naomi Klein, which demonstrates how successful today's neoliberal elite has been in undoing the consensus about Keynesian economics that lasted until the 1970s, and just how patiently they built up their ideology after the Second World War, waiting for the moment when it would be more attractive to voters. So we've now had fifty years of this laissez-faire, free-market capitalism, and the damage done to billions of people has been incalculable.

'That's what we're doing now: creating an idea, a way of building community, freed from neoliberalism's dominant narrative, putting justice at the heart of everything we do, anticipating the moment when this is what voters will demand. We have to make that a reality – we know that some climate breakdown will happen – that's already baked in – and that will entail social breakdown, so we have to be ready to find ways of living together more compassionately.'

Phoebe

'The way things have turned out could make me really angry, but I'm not. I've been reading a lot recently about that aspect of non-violence, and what it means for the way we think about other people. I don't want to sound patronising, but instead of feeling angry I just feel sad for politicians, for all those people who are perpetuating this crisis. They're all part of a completely failed system, and I'm not sure that system can be fixed from the inside. It's as Audre Lorde said: "The master's tools will never dismantle the master's house."'

Ella

Things are so much more clear-cut today than they seemed to be back then. Today, these companies and the wider system that sustains them, are knowingly sacrificing the future of humankind to maximise whatever short-term profit they still can from their oil and gas assets. While the politicians who purport to represent us allow them to get away with it.

In that regard, speaking morally rather than politically, they have become 'evil entities' – 'bad actors' at a scale never seen before in the long history of industrial capitalism. Their boards of directors and senior executives know exactly what the consequences of this continuing 'science denial' will be, So, given what we now know about the governance and strategy of these companies, about the corrupting influence they still have, particularly in terms of undermining the integrity and credibility of the IPCC, and about the true costs of climate breakdown today, how exactly does any international NGO (particularly those larger US-based NGOs) continue to justify their multi-million-dollar partnerships with such bad actors?

DEALING WITH A LABOUR GOVERNMENT

When I talk with young Just Stop Oil activists, or read their blogs, they are very aware of all this. Not necessarily chapter and verse of every single failed COP conference, or every single otherwise inexplicable decision by politicians to protect the interests of the fossil-fuel companies, or even of every single new data point in the climate science – but they absolutely know what's going on. And why it's still going on despite all the evidence of just how high a price we'll pay in the future.

Here in the UK, they also know that our democracy has been weakened by the continuing synergies between the fossil-fuel companies and the short-term needs of our economy. And that's why they feel so suspicious of Keir Starmer's Government. The corruption seems to continue. The fossil-fuel lobbyists are as actively deployed along Labour's

'At Leeds, my degree was in human geography and economics. I read much more about radical politics and direct action. I became contemptuous of people talking about mainstream political solutions without really understanding the nature of the crisis. I tried to set up a new project, Student Rebellion, but there was so much tension between the different organisations involved in climate politics! Many of them were into conventional student activism, targeting the university itself, leafletting, lobbying, a bit of marching. But there wasn't even agreement between those who were more radical, and certainly no agreement on whether we were going to break the law. Not at that stage.'

Sam

'The climate movement is generally too white, too middle class. I imagine that you found lots of "white guilt" in talking with other JSO colleagues? It can create a lot of tension. But for me, it just means we must approach these things differently. It would be crazy to expect everybody in the UK to understand all the details of climate science, or the problems with our food system – especially if you're having to shoplift to feed your children, which many people in Hull are. But we haven't managed to make the connections between the climate crisis and the way so many people are forced to live today.

'I have a lot of these same privileges, which is why I really want to make the most of my academic work. I still think it will be possible to use that to good effect, and I may well return to direct action at some point. But I see the adaptation work of Community Hull as being fundamental – and in no way contradictory to the civil disobedience advocated by JSO and ER. I spend a couple of days a week with my fellow project worker, to pay the rent, then put the rest of my energy into these community projects. That means a pretty frugal way of life, but, collectively, it's incredibly rich.'

Sean

corridors of power as they were with the Conservatives. And when it comes to the question of 'new oil and gas licences in the North Sea', a completely arbitrary line is still being drawn between Labour's commitment not to license any new proposals that might be forthcoming, now that it's in power, and all the new licences issued by the Conservatives but where development has not yet started.

And that latter category includes Rosebank. This is the UK's biggest undeveloped oilfield, 80 miles off the coast of Shetland in the North Atlantic. Back in September 2023, the Conservative Government approved the application from the Norwegian oil company Equinor to develop Rosebank, which is reckoned to have reserves of around 500 million barrels of oil. Burning Rosebank's oil and gas would produce an astonishing 200 million tonnes of CO_2. Consummate liars that they were, Tory Ministers claimed at the time that this was all about lowering fuel bills for 'hard-pressed families' here in the UK. In fact, most of Rosebank's oil will be exported for refining overseas, with a limited amount sold back to the UK at market prices.

In January 2025, the licence granted to Equinor for Rosebank in 2023 was declared unlawful by The Court of Session in Edinburgh on the grounds that Equinor's application, described by Ed Miliband at the time as 'climate vandalism', took no account of all the emissions that would be caused by burning the oil and gas extracted from the field – those 200 million tonnes mentioned above. The judgement did not require Equinor to stop work in developing the field, and the company has confirmed that it will submit a new application taking the Court's ruling into account – in effect, making a new application. This will be by far the biggest test of Labour's integrity in delivering on its net-zero commitment.

It was literally incomprehensible for Just Stop Oil activists that a Labour government might allow such a monstrosity to proceed – and still feel comfortable putting their hands on their ministerial hearts to talk up their decarbonising zeal. Tried and tested climate campaigners are inured to hypocrisy and to Orwellian levels of doublespeak, but Rosebank is in a different league.

'It's possible that Just Stop Oil won't exist by the time this book is published! I don't think that really matters. My own view is that Just Stop Oil should have gone out with a bang – either when the Labour Party agreed that there should be no new oil and gas licences in the North Sea, or when they got elected in July last year. We would have been able to claim a legitimate victory as that was JSO's sole demand at the start. We'd deliberately settled on a winnable demand, with the clear intention that once won, the campaign would end. And that would have made it possible to demonstrate that this kind of non-violent action really works.'

Oliver

'There are still so many people who find it easier to ignore things, because it's not "in our face" right now – even though it's blindingly obvious that it's going to be "in all our faces" pretty soon! They seem to be stuck in a state of false consciousness, with another dopamine fix always in front of them – it really is as if we we're all living in that movie *Don't Look Up*, with a comet that will destroy human civilisation about to hit the Earth.

'I don't spend a great deal of time trying to explain this to people, but when someone speaks to me, I'll talk through the reality of it all. But even my family don't really want to know. The conversation tends to shut down immediately as soon as it's mentioned! And I really don't like preaching about this.'

Harrison

Keir Starmer himself does not have a climate-friendly bone in his body. He's been forthright about his contempt for Just Stop Oil. Labour's Home Secretary, Yvette Cooper, has shown no indication that her Government will consider repealing the draconian laws introduced by the Conservatives, which I deal with in Chapter 8, or even spoken out about the importance of freedom of speech, the right to protest or the independence of juries.

It is only Energy Minister Ed Miliband who has retained any kind of credibility – and even here it's decidedly qualified. There's no doubt he's sincere about pushing as hard as he can for accelerated investment in renewables, with a particular focus on both onshore wind and offshore wind – including more floating offshore wind farms to take advantage of more reliable wind resources.

But his decision in October 2024 to commit £22 billion of public money over the next twenty-five years to Carbon Capture and Storage (CCS) was seen as just one more instance of selling out to the fossil-fuel industry – after the massive lobbying campaign it conducted inside the Department of Energy and Net Zero over the last few years. Using CCS on existing gas-fired power stations is seen by all climate campaigners as a costly boondoggle designed primarily to extend the life of fossil gas; and hugely costly plans to produce 'blue hydrogen', using fossil gas and CCS, are even worse in that this would entail new fossil-fuel investment.

Just Stop Oil's opposition to this abuse of public money used to get right up Keir Starmer's nose. Writing an article in the *Sun* on 3 October 2024, he set up a wholly false stand-off between 'JSO and working people', claiming that 'the jobs of brickies, sparkies and engineers – the backbone of Britain' – will be put at risk by what he called 'Net Zero extremists'. Predicting a 'slow decline back to the Dark Ages' if the extremists were to win that battle, he really let rip: 'I will not sacrifice Great British industry to the drum-banging, finger-wagging Net Zero extremists.'

Keir Starmer is not known for either his sense of humour or his use of irony, but his speech-writers must have been chortling away at having successfully inserted into his article an almost verbatim variation on Mrs Thatcher's grim warning, in a speech at the Royal Society of Arts in 1988, that environmentalists and anti-nuclear campaigners would take the UK back 'to huddling in our caves with an open fire'.

That kind of rhetoric sounded pathetic then. Thirty-six years on, in Starmer's lifeless prose, it sounds even more pathetic.

'You've got to believe that what you're doing is absolutely the right thing. But at the same time, we really want to be successful, so it's important to think about what works and what doesn't work – just so long as you don't go on rationalising about this forever and never do anything! I don't really have much time for people who talk about all this political theory stuff, and then won't put themselves on the line when it comes to it. This is an Emergency, after all!'

Daniel Knorr

DANIEL KNORR

I am twenty-two years old and, until recently, I was a student of biochemistry at the University of Oxford. I grew up in London in a family of five, with an Italian mother, a Croatian father and two younger siblings. I intended to study biochemistry, with a view to engineering plants to be more resilient to extreme weather brought on by the climate crisis, but I quickly realised that I was kidding myself. Five decades of the most brilliant scientists in the world pouring years into research, advising politicians and broadcasting to the public – and yet humanity is still on track to butcher billions of its own via the climate crisis. This is a political problem, and only determined non-violent resistance has a chance of making a real difference.

INVOLVEMENT
August/November 2022: Writing to local MPs/Councillors, door knocking and marching with XR.
January–March 2023: Marching with JSO.
April 2023: Arrested for conspiracy to spray 'Dippy' the Diplodocus, in the Natural History Museum, with orange cornstarch.
May 2023: First person arrested under the 2022 Police, Crime, Sentencing and Courts Act in Parliament Square.

June 2023: Arrested for invading the pitch at The Ashes test match at Lords; carried off by English wicket-keeper Jonny Bairstow.

October 2023: Oxford University paint action coinciding with actions at nine other English Universities.

November/December 2023: Arrested four times slow marching with JSO in London.

February 2024: Arrested in Edinburgh during a Palestine-focused action.

March 2024: Arrested outside Elbit weapons factory in Shenstone for allegedly locking on.

April–July 2024: Months spent assisting Palestine Camp at University of Oxford.

August 2024: Arrested and remanded in prison for conspiring to take action with Just Stop Oil at airports. In May 2025, I was given a custodial sentence of two years.

MOTIVATION

I find answering questions about my motivations quite tricky because taking action at this time just seems so obvious and logical to me. Sometimes I can't think of what else I would be doing. The climate crisis is a genocide on such a scale that it threatens any job prospects I would have in the future, along with the safety of all my loved ones; I don't see the point in doing much else other than trying to reduce the harm. I suppose I do see the opportunity to build a better, kinder and fairer world, and this too drives me on.

INSPIRATION

I'd love to answer this question by naming one of the historical greats like Angela Davis or Nelson Mandela – it would make me sound very well read! But if I'm being honest, it's my closest friends who inspire me the most. I've had the fortune of hearing my good friend Cressie Gethin speak on several occasions. She never fails to amaze me with her honesty, determination and vulnerability. I remember attending a talk by Chris Packham; Cressie was speaking at the end and I remember hearing her speak and thinking that she had completely blown Packham away. Her speech moved me to tears. Cressie is probably my biggest inspiration.

IN NATURE
My mother grew up in a rural part of southern Italy. Her dad was a farmer and co-owned a small bit of land up in the mountains. As a child I'd go up there every summer. There was always so much life. My grandad kept chickens and pigs, the farm was teeming with grasshoppers and ants and other bugs, and there was a pond down in the valley with trout swimming about. There'd be wild fennel and blackberries everywhere so you could stuff your face while you were running about. I haven't been for a few years, but I'm sure if I went back, I'd feel right at home.

QUOTATION
In the prison chapel they have a framed quote by Martin Luther King: 'Hate cannot drive out hate, only love can drive out hate', which I have been enjoying every time I go to the chapel.

RESOURCE
I do think the podcast Designing the Revolution by Roger Hallam is required listening at this point. I also really enjoyed the podcast Revolutions by Mike Duncan. It's great if you want an overview of Western social history starting with the English Civil War and finishing with the Russian Revolution.

WHAT LIES AHEAD?
I don't believe in utopias, but I do hope that at the very least there will be a system in place which will put some form of human well-being above ever-increasing profits. I would hope to see ordinary people involved in decision-making, with enough engagement at a local level all the way up to the top so that anyone can feel they will be involved in some form of decision-making at least once or twice in their life. I don't think at this point we can stop much of the damage from the climate crisis, so a system that listens to all voices and treats people with compassion is essential if society is to adapt and successfully survive the gauntlet.

EDDIE WHITTINGHAM

I grew up in Cambridge and I'm currently living in Exeter. I graduated from Exeter University with a BA in politics, philosophy and economics, but I was a campaigner before and during my uni years.

INVOLVEMENT
I was previously an active supporter of Extinction Rebellion, with my first ever protest being the April 2019 Rebellion in London. Since 2022, I've been an active supporter of Just Stop Oil, and I'm currently also involved in Youth Demand and other revolutionary campaigns. I've been arrested ten times. I've been found guilty three times in Magistrates' Courts, and I've been to prison twice on remand. I disrupted the Snooker World Championships in April 2023 and, to my surprise, ended up on several front pages, with the action becoming a global news story. In April 2024, I was found guilty of criminal damage to the Treasury. I received a suspended sentence and have to do 100 hours of unpaid work and pay £500 in compensation.

MOTIVATION
What drove me to take action to the point of arrest and imprisonment was, broadly, two things: first, it's clear to anyone with even a passing interest in the climate crisis that the decades of diplomacy and conventional campaigning have utterly failed to bring about the change we need; second, we face an intolerably horrific future of mass starvation, mass migration, war and death as a direct and indirect result of the climate crisis. I think it's entirely possible

that I won't live a full life, and I definitely won't be having children as I don't believe there will be a safe enough world for them to grow up in.

INSPIRATION
My main inspiration for taking direct action has been a man called Roger Hallam. He was a peace activist in his youth who went on to become an organic farmer in Wales. His crops were destroyed by increasingly erratic weather, and this prompted him to study a PhD in how to bring about rapid social change. He managed to get King's College in London to divest from fossil fuels by taking direct action, including a hunger strike. He then went on to co-found Extinction Rebellion, Insulate Britain and Just Stop Oil. He is one of the few people who I'm in awe of, as he embodies a spirit of genuine selflessness, service, truth-telling and radicalism.

RESOURCES
Advice to Young People as They Face Annihilation, by Roger Hallam. Available on YouTube.

WHAT LIES AHEAD?
I don't have hope that we will sort out the climate crisis in anything resembling a proportionate or just way. However, I do believe that a spiritual revolution – a fundamental shift in how we see life – is entirely possible and even plausible, as it is likely to be forced on us. The full extent of my hope is that as many humans survive as conditions permit, and that they may live in a more fulfilled way.

We are the stories that we tell ourselves, and we have to start telling better stories! The other part of my life is the work that I do as a writer, and this is where I think a lot about the insights of Amitai Etzioni and his idea of the Common Good. There are really just three things that make us happy and fulfilled: the quality of our relationships; the quality of our intellectual and spiritual life; and the connections we have with community.

EILIDH MCFADDEN

I grew up in West Lothian and currently live in Fort William, studying adventure education at the University of the Highlands and Islands whilst also working on a climate and cost-of-living campaign in Scotland called This is Rigged. Any spare time is usually spent painting walls on community service, climbing hills or crocheting bright pink balaclavas.

INVOLVEMENT
I joined Just Stop Oil in January 2022 and took part in actions like tanker surfing, climbing onto the pipework of oil terminals, breaking injunctions, blocking roads, painting government buildings, slow marching and throwing cake at the waxwork of King Charles in Madame Tussauds.

During that time I also took part in actions with Animal Rising – drilling tyres – and with a social-housing campaign in Glasgow trying to stop the destruction of flats.

In January 2023, I started focusing my time and energy on This is Rigged. I have been arrested for shutting down the oil refinery in Grangemouth and for spray painting.

MOTIVATION
What motivates me is knowing that there are so many people before me who have taken part in direct action throughout history, fighting for a better world, and there will be many more after me. Some of the people and movements that I look to for inspiration haven't always won their demands, but their work has put pressure on those they are challenging and paved the way for those who

come after to win. It can help to make you feel less isolated, and gives me hope in times when progress doesn't feel as tangible.

INSPIRATION
Abbie Hoffman, a political activist in America during the 1970s, who co-founded the Yippie movement (the Youth International Party, YIP), was a defendant in the trial of the Chicago 7, and later went on to write Steal This Book.

IN NATURE
The river Avon next to Muiravonside in Falkirk. When I was a wee child, I used to walk there every weekend with my family to visit my favourite tree, a yew that I had named 'elephant tree' because of its branches that resembled an elephant's trunk and tusks.

QUOTATION
'There is a time when the operation of the machine becomes so odious, makes you so sick at heart, that you can't take part. You can't even passively take part. And you've got to put your bodies upon the gears and upon the wheels ... upon the levers, upon all the apparatus, and you've got to make it stop And you've got to indicate to the people who run it, to the people who own it, that unless you're free, the machine will be prevented from working at all.'

Mario Savio, 1964

RESOURCES
Blueprint for Revolution: How to Use Rice Pudding, Lego Men, and Other Nonviolent Techniques to Galvanize Communities, Overthrow Dictators, or Simply Change the World, by Srđja Popović.

WHAT LIES AHEAD?
Obviously, a win would be to see our government act like the situation is as serious as it is, and to finally start listening to the scientists and experts. But on a deeper level than that, success to me looks like seeing a shift to greater public awareness on the power normal people have to make change.

'Much of my early campaigning was at uni. I led a campaign against the £15-million partnership the uni had with Shell, despite our university having some of the best climate scientists in the world! The 'Shell Out' campaign really angered the Vice-Chancellor and I was astonished to find myself on the receiving end of intense hostility. They used every weapon they had at their disposal to get me to step down – I had become President of the Union, a full-time, sabbatical role – even though our protests were all perfectly legal.

'I learned a lot about past student protests: the anti-apartheid campaign in 1968 and various decolonisation protests. My uni had been in the forefront of the campaign against Barclays Bank and its support for apartheid in South Africa. So I was disappointed that so many students didn't want to get involved in the Shell Out campaign.

'But I couldn't really blame individuals. Ever since Margaret Thatcher, universities have been turned into capitalist institutions and students have been largely depoliticised. Student loans and increased costs mean that more and more students are focused on "climbing the ladder", doing what they need to do to ensure a good job, and finding as much part-time work as possible during vacations.'

Emma

'It is a bit strange that so many people of our age can still stand by and hope that something will just "happen" to avoid the otherwise inevitable collapse. But what really pisses me off is not them being ignorant or indifferent, but the systems of oppression that keep them ignorant and indifferent. And the media that perpetuate this. On a scale of 1 to 10, in terms of being pissed off, it's definitely a 10 when it comes to the corporate media and all the fossil-fuel lobbying!'

Paul

5: THE FOSSIL-FUEL INCUMBENCY

'Climate activists are sometimes depicted as dangerous radicals. But the truly dangerous radicals are the countries that are increasing the production of fossil fuels. Investing in new fossil fuel infrastructure is moral and economic madness.'
António Guterres, April 2023

IN OCTOBER 2024, BP announced that it would be dropping its target to reduce oil output by 25 per cent by 2030. The original target, set by the then CEO Bernard Looney back in 2020, had been to cut output by 40 per cent by 2030, as part of a much wider transition strategy, to reduce dependence on fossil fuels, but that was then scaled back to 25 per cent in 2023. And then Murray Auchincloss, Looney's successor as CEO, got rid of the target altogether.

As ever, 'market pressures' were given as the reason. BP had made record profits of $28 billion in 2022, and shareholders were unhappy that they would be missing out on equally lucrative paydays in the future if production was going to be cut. Auchincloss set out to reassure them, not only by getting rid of the target, but by announcing huge new investments in Iraq and the Gulf of Mexico. Who needs a transition strategy when the big bucks keep rolling in, and governments remain as clearly 'addicted' to fossil fuels as they've always been?

And then, the coup de grâce in February 2025, when Auchincloss announced further swingeing reductions in investment in clean and renewable energy in order to ramp up fossil fuel investments to $10 billion a year. At least he had the good grace to acknowledge that nothing else now matters to BP apart from its shareholders: 'Our optimism for a fast energy transition was misplaced, and we went too far, too fast. This is now a reset BP, with an unwavering focus on growing long-term shareholder value.'

'I don't get mad at the fossil-fuel companies themselves. I'm mostly mad at the environmentalists, who knew about all this stuff thirty or forty years ago, and just let things go on getting worse and worse. Now it's our generation that has to do the heavy lifting. Even today, a lot of those people who let this happen do nothing to disturb the nice lives they're still attached to.

'For example, they had achieved literally nothing in terms of putting an end to new licences for oil and gas in the North Sea, then Just Stop Oil made that happen in a few months. OK, only a partial win, but still a win! And I could hardly believe the animosity of mainstream environmental NGOs to the idea that a tiny number of people had achieved what they had failed to achieve, pushing back against the fossil-fuel companies, over so many years.'

Chiara

'This is Rigged is campaigning on what a just transition looks like in practice, for those currently working in the oil and gas industry. I grew up about ten minutes from the Grangemouth refinery and I know how massively important it is in terms of jobs – even the hope that it gave young people of a good career. Early on, This is Rigged closed Grangemouth for a week, but the planned closure really isn't going to help the climate at all. If anything, it will be even worse, as the oil coming from the North Sea will be refined elsewhere then imported back into Scotland.

'So, for us, this is all about workers' rights, fair pay, good working conditions, being able to afford the basics in life – all these things really matter, particularly in a small community like Grangemouth.'

Eilidh

BP is no different from the rest of today's Independent Oil Companies (IOCs), but it's 'back and forth' strategy over the past twenty years (from 'beyond petroleum' back to 'fossil fuels forever') tells us a lot about how deeply entrenched is that addiction to fossil fuels.

The easiest way to understand the astonishing reach of the fossil-fuel incumbency is to see it as a global imperial power, operating in every corner of the Earth, regardless of the political status of countries, whether democracies, autocracies or failing states, subject only to partial and ineffective regulation by those countries once they've been effectively 'captured'. This is achieved by the limitless amounts of money and other inducements the industry has deployed throughout that time to persuade politicians where their best interests lie. Equally limitless amounts of money are available for marketing and advertising campaigns of every description, for sponsorship arrangements and for high-profile charitable activities.

All of this is possible for two principal reasons. The oil and gas companies that have dominated the global economy for the last thirty-five years, and the coal companies who did so for a lot longer than that, have been extraordinarily profitable. In 2022, analysts at the University of Antwerp in Belgium calculated that the average profit generated by the oil and gas industry, including both Independent Oil Companies and National Oil Companies, amounted to a staggering $1 trillion a year over the fifty years between 1970 and 2020 – that's $2.8 billion a day in pure profit. As Professor Verbruggen, the lead author of the report, said at the time: 'You can buy every politician, every system, with all this money, and that's what's happened. It protects these companies from political interference that may limit their activities.'

Every year, a share of those profits is paid out as dividends to well-satisfied shareholders; a share is reinvested in new assets – in exploration, development and production; and a share goes back to governments and their citizens as tax revenues on the licences issued to those companies. But that has still left an astonishing amount of money sloshing around for other nefarious purposes.

'I'm one of the few people in my circle who is so vocal about how angry I am! Just Stop Oil is a non-violent movement, so there's a lot of talk about redirecting one's anger, that "this isn't the time to be angry, it's the time to be helpful and in service to others". I was never very good at that; I can't help but feel overcome by the injustice of it all, particularly when I think about fossil fuels, that those companies knew what would happen to the climate way back in the 1960s. So they've been preparing for this future for decades, since before my parents were born! And we're the ones who now have to do the fighting. I feel so infuriated that it's my generation that is going to be faced with all these consequences.'

Anna

'We're all very aware that a lot of people aren't in a position to take the same action that we have. Ninety-six per cent of students studying petrochemical engineering at Birmingham University end up working with the fossil-fuel giants! A lot would say they don't really have a choice, and it's not really about the individuals here, it's about the companies. They might still be making a wrong choice, but there are so many bigger forces at work.

'It's those companies that make me angry for the way in which their actions take away people's choices. When you know that things are going to keep getting worse, inevitably, there are such huge consequences. I wouldn't have taken the action I did if I didn't feel that kind of anger.

'But it's more the love I feel for the people whose lives are devastated by these companies, than anger for the companies themselves. I made a decision some time ago that it's just not possible to have children in such a world – it's not a compassionate thing to do, inflicting that kind of reality on children in the future.'

Harrison

The second reason is that none of these companies has ever, at any stage in their history, been required to pay for the social and environmental costs incurred in bringing their products to market. Governments have simply permitted them to 'externalise' the cost of all those billions of tonnes of greenhouse gases released into the atmosphere. That doesn't mean those costs disappear: it means that they're paid by individuals and communities affected by their often grotesque polluting activities, by the environment – in the form of pollution of soil, water and forests – and, of course, by future generations.

EXTERNALISED COSTS

Over the years, economists have gone to great lengths to calculate the scale of those externalised costs, particularly the emissions of greenhouse gases. In November 2022, the US Environmental Protection Agency assessed the cost of each tonne of CO_2 emitted at around $200, but an influential paper published in May 2024 assessed the full social cost to be over $1,000 a tonne. Other academics and policy experts have come up with figures varying between $50 and $1,000.

From the perspective of young people today, all this talk of 'externalised costs' sounds very academic. The simple reality is that people alive today are all – collectively – enjoying the benefits derived from the use of those fossil fuels, just as previous generations have done. Until now, however, people alive today have paid for none of the costs associated with the emission of greenhouse gases – primarily CO_2 and methane – arising from their use. In effect, we've been systematically subsidising ourselves year after year after year.

However, more and more of those costs are now showing up in real time, in the form of economic damage done by climate-induced disasters and extreme weather conditions. In Florida, in 2024, around 450,000 claims were filed arising from the impact of Hurricanes Debbie, Helene

'In my third year at uni, I started studying ecology and environmental science. It's harrowing learning all the things that are going wrong with the world, and why everything is falling apart. So I also read through the annual reports of BP, Shell and other fossil-fuel giants, checking out what their spending was on green and fossil-fuel energy. In stark contrast to their adverts, these companies were spending orders of magnitude more on fossil fuels than on green energy. I almost obsessively went back over the numbers, again and again, finding it really difficult to believe that people at these companies were planning to inflict on us the worst-case scenarios I'd been learning about in my university studies.

'My first action was disrupting a session organised by Shell for geophysics masters students at Cambridge. I can remember feeling quite nervous, not seeing myself as the kind of person who wanted to break the rules! But listening to that guy from Shell made me boil with rage, especially when he shared with the students how "immensely proud" he was of all the work that Shell had done in Nigeria! We'd had direct contact with Ogoni activists in the Niger Delta and had heard horrendous accounts of the devastation caused by Shell's activities, year after year.

'But even when I think about people working for Shell, including those who are somehow proud of what they've done in Nigeria, I feel disdain rather than rage. How can they not be asking themselves the critical questions? How can they sleep at night? How can they live with themselves?'

Sean

5: THE FOSSIL-FUEL INCUMBENCY

and Milton, with estimated insured losses of $5.5 billion. That sounds like a lot, but Hurricane Ian in 2022 caused losses of an astonishing $21 billion, driving up insurance rates across Florida through 2023 and 2024. And those are the insured losses; on top of that, there are the under-insured losses and – most traumatic for individual homeowners and businesses – the uninsured losses. Numbers start rising dramatically at this point, as pointed out by AccuWeather's Chief Meteorologist, Jonathan Porter, referring to the USA as a whole:

> This has been a tremendously expensive and devastating hurricane season. It will be remembered for shattering records and causing approximately $500 billion in total damage and economic loss. To get this in perspective, this would equate to nearly 2% of the nation's GDP.

Let's just pause for a moment here. Half a trillion dollars. Two per cent of national GDP. From one hurricane season. In one country.

Since then, the economic damage caused by the devastating wildfires in Los Angeles in January 2025, with at least twenty-nine deaths and 16,000 homes destroyed, has been assessed as a minimum of $150 billion, with far higher under-insured and uninsured losses on top of that.

As far as I can tell, there is currently no single aggregated calculation of economic losses arising from all climate-induced disasters and extreme weather events in all affected countries that occur in any one year. But I hope your imagination is grappling with this unthinkable calculus of economic loss and personal suffering.

And then push your imagination a little further, bearing in mind that this is the 'bill' arising from an average temperature increase of no more than 1.3°C since the time of the Industrial Revolution, and extrapolate from there into the terra nullius of all the bills that await us at 2°C or even 3°C. Or, to be more accurate, await the citizens of the future: the young people of today.

'I can't pretend I'm OK about all this, and I do get angry about the continuing failure to see the climate crisis for what it really is – simply because this robs everyone I love, and literally everyone on the planet, of a safe and decent future. So I've ended up "weaponising" my rage, especially when it comes to the fossil-fuel companies and the way they play so dirty. The way I see it, politicians today don't have the freedom or the power to control those companies, even if they wanted to, in order to end our dependence on fossil fuels. It genuinely makes me want to tear my hair out!'

Hanan

'On some days I feel a burning rage about all this! But I know that anger is not particularly helpful – it just eats away at me. Even when I'm thinking about the fossil-fuel companies, I don't think anyone's an inherently bad person; they end up doing bad things because they are hurting, or caught up in a system which is so damaging, or trapped within a model of what success looks like and don't consider they have any choice. I think that's just sad, especially when it leaves them so detached from human suffering that they don't see any need to change.'

Olive

'We all see things slightly differently, and there are some colleagues of mine who think that we are up against really evil people, particularly in the fossil-fuel companies, who continue to do really bad things even though they know exactly where we're going to end up. They're the bad guys, and we're the good guys! I know that narrative can sometimes be useful on social media, but I don't really feel it does justice to the nature of the challenge we face. Sure, lots of people seem to go on making really bad decisions, including those who seem perfectly happy to work for a company like BP, but does that make them evil? I don't think so.'

George

5: THE FOSSIL-FUEL INCUMBENCY

I'll return to this in Chapter 9, but this is such a shocking example of Intergenerational Injustice that it's hard to believe the level of invective young climate campaigners are subjected to simply for trying to get today's 'grown-ups' to start paying a bit more attention. Any suggestion that the industries primarily responsible for these current and future bills should now be held to account, both politically and financially, is still peremptorily dismissed as unworldly or, worse yet, as prejudicial to shareholder interests and to capitalism itself.

There's never been an incumbency as pervasive and powerful as this one. It's not just the companies themselves, comprehensively dominating the visible foreground, that make up this incumbency, but just behind the scenes there is an even more extensive network of financial and professional interests that provides the funding; facilities; insurance, legal and consultancy services; and the vast array of transport, infrastructure, logistic and retail businesses that distribute and sell the industry's products.

There is one final aspect of this incumbency that has shielded it from what would otherwise have been a much more rigorous scrutiny of the balance of costs and benefits generated by it: the media. Even as the power of the industry grew in the second half of the last century, large media conglomerates began to dominate the global scene, ever more concentrated in the hands of a few billionaire owners – both in print and broadcast and, more recently, digital media.

Pre-eminent amongst these has been Rupert Murdoch. In the USA, Australia and the UK, he has used that power to deny and disparage climate science, to trash the reputation of individual climate scientists, to promote the interests of fossil-fuel companies at every turn, to intimidate politicians, viciously attack climate protesters, and deceive readers, listeners and consumers as to the true state of affairs around climate breakdown – in effect, to assume the role of cheerleader-in-chief for the single most destructive force on planet Earth today.

That lack of media scrutiny, let alone anything even vaguely resembling proper 'truth-telling' about the climate crisis, has already cost us very dear indeed. Not least in terms of the trillions of dollars of taxpayers' money that continue to reward the whole fossil-fuel incumbency through direct and indirect subsidies. For those who already know about this utterly surreal situation of governments all around the world enthusiastically deploying tax revenues to accelerate climate breakdown, I'll be brief. For those who don't, this may come as a surprise.

POLITICAL CAPTURE AND CORRUPTION

It's generally accepted that global direct subsidies for coal, oil and gas amount to roughly $500 to $600 billion a year. This includes both production subsidies for the industry itself and direct support for consumers by keeping the cost of fuel below market prices. But beyond that are all the externalised costs for which companies are not being held responsible, coming in at anywhere between $4 trillion and $5 trillion a year. The International Monetary Fund keeps a very close eye on this, providing authoritative data that I'm sure would be dismissed if it came from any other source, and provides a regular update, with more granular, country-by-country detail. For instance, the US Federal Government in 2022 stumped up a quite extraordinary $757 billion in direct and indirect support for fossil fuels. Crunch the numbers, and that amounts to $2,226 per US citizen!

How can one interpret this as anything other than a conspiracy against the American people by their own Government?

Subsidies, direct and indirect, both big and small, remain a powerful tool in government energy strategies. In the previous chapter, I shared a classic example of PCC (Political Capture and Corruption) with poor old Ed Miliband being scammed into stumping up billions of pounds for new Carbon Capture and Storage (CCS) schemes. This is a technology that's been around for a very long time, with an operating record of capturing emissions of CO_2 from the flue gases of power stations, and then sequestering that CO_2 in old oil and gas reservoirs, that is absolutely abysmal. CCS costs are significant; the capture technology itself is extremely energy-intensive; capture rates – the percentage of CO_2 capture – have never achieved the promised efficiencies; and almost all the storage/sequestration achieved so far has been to re-inject the captured CO_2 back into declining oil and gas assets to squeeze even more oil and gas out of them!

Just a quick reminder here: this is a Government that has a target of 2030 to achieve a 95 per cent decarbonised electricity supply system. To achieve this, it will need to phase out most of our existing unabated gas-fired power stations, i.e., those that don't capture and store CO_2 emissions. Confusingly, however, none of these has been prioritised for retrofitting CCS, which would theoretically extend their generating lifespan. Instead, a brand-new, £4-billion, gas-fired plant sits at the top of the CCS prioritisation table. Courtesy of a £1.4 billion bung from the British Government, BP and Equinor are promising to build one of the biggest hydrogen plants in the world (using fossil gas and CCS) to be operating as a 'completely Net Zero asset by 2030'.

This is such a desperate lie as to make even Boris Johnson blush. Even if a huge Gas + CCS plant could be successfully installed before 2030, and even if it achieved the promised capture rate of 95 per cent, and even if all that CO_2 was then safely sequestered, via a completely new pipeline, in spent oil and gas reservoirs in the North Sea, it still wouldn't be a net zero-asset. And that's because huge amounts of methane, so-called 'fugitive emissions', will have been unavoidably released in the upstream business of extracting and transporting the gas to be burned in the plant.

The sheer scale of this boondoggle is staggering. But from BP's point of view, it's a fantastic prospect – not because it's ever likely to be hugely profitable, but because that £1.4 billion of UK taxpayers' money serves as an excellent proxy for genuinely earned profits. BP shareholders really couldn't care less where their dividends come from. But all the rest of us should.

ALEX DE KONING

I never saw myself as an activist. It wasn't long ago that I was walking through campus with my headphones on when someone gave me a flyer. It read, 'Find out what the young are doing about climate change.' I'd always been concerned about it – otherwise I wouldn't have chosen this specific PhD in green hydrogen production – so I decided to go along. My closed-off mind felt like it had been smashed open by a sledgehammer.

Two weeks later, I ran in front of a moving oil tanker, sat down on the road, and waited for fourteen hours to get arrested. In the following three weeks, I was arrested twice, took a week-long holiday off work and spent nine days in London participating in as many actions and marches as I could. I'd never been to a protest of any kind before this.

And now, here I was, grabbing my gear and sprinting towards the gates of the Nustar Clydebank oil terminal in Glasgow. I propped the stepladder up against the fence and we scrambled over, one by one. Once on the platform, we then had the far more difficult job of getting onto the pipes above us. I did a lot of rock climbing when I was younger, so I volunteered to be the first to try and find a safe route for the others to scurry up onto the pipes.

When we reached the top, there was a sense of contagious ecstasy: not only had we succeeded in getting onto the pipes when other groups in the past had been tackled by security, but also – because the pipes were wide, at the same height and had only small gaps between them – we'd be able to find a place to sleep. But two of my good friends never made it onto the pipes – I had no idea that they were scared of heights and was amazed they'd even volunteered for such an action.

We settled in and the first few hours seemed to fly by, mostly watching the security guards watching us. The police came and asked us all the standard questions that we were used to. 'No, officer, we have no intention of coming down any time soon.' After some discussion about how many Imodium is too many, and why we hadn't packed more than two pee bottles, it was time for bed. I'd wedged myself securely between two of the pipes, so I felt safe going to sleep. However, waking up, forgetting where I was, and accidentally looking down to the sheer life-threatening drop on my right-hand side, gave me a jolt that made it impossible to sleep afterwards. It was the most terrifying moment of my life.

It wasn't long before the police officers returned, marching towards us with renewed purpose. My two friends on the platform below were immediately pushed against the railings and aggressively read their rights. The officers wasted no time in getting their specialised heights team out, and securing my friends in harnesses. Once strapped in, one of them looked up at us, saw the phone camera was pointed at her, and started giving a testimony like I had never heard before.

She'd given the very first talk I went to. I wouldn't have been here without her. She was an environmental scientist who'd started working for an environmental charity. A few months in, she filled in a feedback form saying that this charity needed to push further and faster and that they were dangerously close to greenwashing. The next day, she was fired. Then she joined Just Stop Oil. And now she was crying – crying because she was scared of heights, crying because she was being manhandled, crying because the weight of the climate crisis was pressing down on her, crying because she was disappointed that she couldn't block oil for longer. I had to look away or I would have ended up in tears myself.

The two in my group were both ready to go down. They looked exhausted. I was trying to work out how to lock on just by myself, but the pipes were simply too wide. I tried lying down on a pipe, and wrapping my hands around it to be glued underneath, but all that did was to crush the phone on which all those testimonies had been recorded before, but hadn't yet been uploaded. So no one else would ever see my friend talking about the injustice of the climate crisis while being forced, crying, to abseil down an oil platform. Out of everything I'd been through over the last two days, this was the worst.

After many hours of waiting in the police van, being transported to the police station and all the paperwork of being checked in, I went straight to sleep. As soon as I got out the next morning, I checked the news for any sign of our action. Despite a handful of local news articles, there was no mention of it in any of the larger UK outlets. I scoured the internet for the interview I gave the morning of the action, but I needn't have wasted my time.

Yet again, the mainstream media had somehow come to a collective decision to completely ignore us.

CRUSHING THE INCUMBENCY

So what exactly is needed to ensure that the fossil-fuel incumbency is properly put to the sword?

1. End All Subsidies
Ever since 2009, meetings of both G7 and G20 countries have committed to do exactly that. Very little has happened. Massive public pressure is now needed.

2. End Foreign Investment In Fossil Fuels
One formidable part of the incumbency has been to secure multi-billion-dollar investments in new fossil-fuel assets from international and regional development banks. The International Finance Corporation (IFC), the Asian Development Bank, the European Bank for Reconstruction and Development, the Japanese Development Bank and others are still only too happy to finance such assets, in the full knowledge that although they would usually be expected to operate profitably for decades, that simply won't be possible in a net-zero world. This means that at some point between now and 2050 they will all become stranded assets. So why, one has to ask, did investment in new fossil fuels via development banks increase from $1.2 billion in 2021 to $5.4 billion in 2022? This is insane.

3. Kill Coal
This can only happen via the kind of disruptive market forces that have been unleashed in the past. As prices for all renewables continue to fall, and efficiency and reliability continue to improve, almost all energy commentators now see 'the renewables revolution' as completely unstoppable.

But there's so much more that could be done to drive this revolution even more aggressively. One massive barrier is the cost of capital for investing in renewables in developing and emerging countries: according to the International Energy Agency, the cost of capital for renewables in such countries is still more than twice as high as for fossil fuels. This, too, is insane.

4. Wind Down Oil

We need to go from a current demand level of around 95 million barrels of oil a day to less than 5 million barrels a day in the shortest possible period of time.

That will rely on two huge market shifts: first, accelerating the transition out of petrol and diesel vehicles into electric vehicles (EVs). Even those who were once deeply sceptical about EVs now grudgingly acknowledge that the internal combustion engine is on death row – it's only a question of how long it will take to carry out the execution. That timetable is being driven almost single-handedly by the Chinese for geopolitical and market domination purposes all of their own.

Second, we have to restrict any further growth in plastics, this being the one market in which oil producers and refineries still believe they have decades of profitable expansion ahead of them. This is now the biggest and most important battleground in our war against the fossil-fuel incumbency. And it is a war. Negotiations around the Global Plastics Treaty in South Korea at the end of 2024 were totally derailed by fossil-fuel companies refusing to accept the need for a mandatory cap on plastics production, arguing instead for investment in downstream waste management and recycling – which it knows only too well has been stuck at less than ten per cent for many years.

5. Throttle Gas

This is not so easy! Huge investments are still pouring into new gas-fired power stations. Liquified Natural Gas (LNG) production and export, from the Middle East and the USA in particular, is booming.

Economists argue that the only way of halting this growth, and then winding the use of gas right back down again, is via the price mechanism. And the only truly effective way of doing this is going to be through a global carbon tax that will apply as much to methane as it will to CO_2 – in other words, strategic, globally-enforced 'cost internalisation', to use the geeky economic jargon. And that means gradually and consistently ratcheting up the cost per tonne of CO_2 until something resembling 'a level playing field' between the use of gas and alternative energy technologies emerges.

Until now, it's been impossible to imagine a political context in which world leaders would do what needs to be done here. The sheer fire-power of the fossil-fuel incumbency has always overwhelmed the combined advocacy of climate science, NGOs and ordinary citizens and voters. I believe that is about to change. And the improbable new element involved in that is the insurance industry.

The global financial system depends on three vast sub-systems: banking, which remains as morally bankrupt today and as incapable of doing anything other than maximising short-term profits as it has always been; investment and asset management, now fitfully shifting a little as the risks of stranded fossil-fuel assets and potentially draconian government regulation loom ever larger; and insurance.

I've allowed myself a little bit of applied day-dreaming (see p. 250) to show how the near collapse of the reinsurance industry, in a rapid and utterly unprecedented socio-economic tipping point, precipitates a financial crisis which makes the Great Financial Crash of 2008 look like a minor blip in the world's capital markets. Such a crisis will be horrendous, and the lives of huge numbers of totally innocent people will be devastated. But much, much worse damage will be averted in the process, as it brings about a just-in-time 'course correction' in the global economy.

ELLA WARD

My mum and dad brought me up with strong moral convictions. They told me to ask questions. They always encouraged me to do the right thing. They showed me what living my life not just for myself, but for others, looks like. They've been there for me through everything, stood by me and supported me, and constantly inspire me.

I've always been interested in climate change, not least as I was obsessed about clouds and ice! Which I guess is why I decided to study environmental science at Leeds. I only started to wake up to the crisis in my second year, with a module on 'Climate Change Science: Science and Impacts'. I later went to a JSO meeting, and this was the first time I had properly connected emotionally – with what's going on.

INVOLVEMENT

I'm currently in prison awaiting trial on a charge that carries a maximum ten-year sentence. In 2026, I have two more Crown Court trials for other actions taken with Just Stop Oil. I'm OK with that. Making change requires people to take action, and as state repression increases, it inevitably leads to some of us sitting in prisons for a little while – or even a long while! I have no intention of packing it all in just because I've spent some time in prison.

MOTIVATION

I act because the alternative is to not act – and that's not something I can do. Action is a more logical decision than inaction. When the lives of millions of people are at risk, choosing to do absolutely nothing, and carrying on like nothing is happening, is not just illogical – it's insane! Doing something effective to tackle this crisis is the only logical and reasonable response to the information we have. I have chosen to not be a bystander in the face of massive harm, death and destruction. I can't let this injustice unfold, and close my eyes, look away, pretending it's not real. I can't stand by and do nothing.

INSPIRATION

I'm so inspired by the people of Palestine, who have shown – and continue to show – incredible strength and resilience over decades and generations of colonial rule, occupation, land theft, apartheid, administrative detention, settler violence and war.

I am also inspired by a lot of the women I meet in prison, who have been through impossibly difficult situations and felt the full, often unjust force of a violent, dehumanising, uncaring justice and prison system. But they still act with care and compassion and continue to look out for each other.

IN NATURE

I feel at home whenever I look up at the sky and can watch the clouds. I could be in the most beautiful part of the world, or I could be looking through the bars of a prison cell – but if I can see the sky, and see the clouds, then I am at peace, and at home.

WHAT LIES AHEAD?

There's a realistic chance that I will still be in prison, on tag, or on licence over the next couple of years. But I'll still be putting my energies into whatever I think is the most effective way of making change, whether that is with Just Stop Oil, some other iteration of a Non-Violent Direct Action group tackling the climate crisis, or something else entirely. I'm open to all possibilities!

I think everyone will be more aware of the climate crisis by then, because we'll all be living with the devastating physical consequences of it, with more disasters happening across the world, and more extreme weather here at home. I think the suffering will be worse for so many people.

I hope I'm wrong. I hope the government commits to stopping extracting and burning fossil fuels as soon as possible. I hope we'll be working with other governments internationally to respond to this global crisis with a united response, including paying reparations to countries, nations and indigenous communities in the Global South that are most affected by climate breakdown and have contributed the least to global emissions.

I hope we're living in an actual democracy, where non-violent protesters aren't surveilled, repressed, criminalised, demonised by the media, and imprisoned by the state. I hope the anti-democratic, anti-protest laws have been revoked. And I hope we live in a society that cares more about people than profit.

(In May 2025, Ella was sentenced to two years for conspiring to cause a public nuisance at Manchester Airport. See pp. 8 & 9 for her letter from prison on her 22nd birthday.)

EMMA DE SARAM

I'm a climate and health justice activist with experience in grassroots organising, Non-Violent Direct Action, social-media engagement, and community-led campaigns. All this started with Fridays for Future in Exeter at uni, I helped set up the Plant-Based Universities campaign, and was elected as Vice President for Liberation and Equality where I led the successful £2 meal campaign amidst the impact the cost-of-living crisis was having on students. I was subsequently elected as President of the Student Union, where we challenged the University's £14 million deal with Shell through our Shell Out campaign.

After attending COP26, I set up a repair café in my hometown with my mum, as a tangible antidote to the despair of sitting in such depressing conference spaces. I give a lot of workshops and talks for broad audiences around climate activism and accessibility, taking an intersectional approach to climate action, demonstrated through my work with trade unions, fostering global solidarity and using my experience with long-term health conditions to make climate action possible for anyone.

INVOLVEMENT
Extinction Rebellion – local involvement and protests/marches.
Just Stop Oil – slow marching over three months, spokesperson, risking arrest on multiple occasions, but not arrested, until, due to health conditions, I was no longer able to participate.
Plant-Based Universities – Co-ordinator for Exeter University.
Green New Deal Rising – action taker, challenging MPs.
Shell Out – lead campaigner.
Exeter Trade Union Council Green Rep.
SOS UK – student trustee.

MOTIVATION
My hope is that there is still a small window of time to slow down climate collapse and build resilient communities and systems so that the most vulnerable people in society are protected.

INSPIRATION
There isn't one individual who inspires me, but groups of people who have collectively taken action; the Civil Rights Movement, anti-colonial movements, the students of 1968, the Suffragettes, Palestine Solidarity marches; throughout history there have always been groups of people who have decided

PERSONAL PROFILES

to challenge the status quo, despite the consequences they knew they would face.

IN NATURE
Anywhere I am surrounded by trees, I can feel at home and grounded in familiarity.

QUOTATION
'Another world is not only possible, she is on her way. On a quiet day, I can hear her breathing.'
Arundhati Roy

RESOURCES
It's Not Just You, by Tori Tsui.

WHAT LIES AHEAD?
I hope that we can slow down what we are told is inevitable, that we have communities strong enough to help each other in disaster, that there are spaces we can go to learn, help out and maybe even live collectively. I hope that climate prisoners no longer exist, that the law, media and politicians see sense, and certainly I hope that Elon Musk, Donald Trump and all their cronies get stuck on Mars and we can keep all their money to spend on global healthcare and climate.

(See pp. 210 & 211 for Emma's article on climate justice and neurodivergence.)

LOVE, ANGER & BETRAYAL

GEORGE SIMONSON

I studied mechanical engineering at the University of Edinburgh, where I found my niche in researching novel recycling technologies to deal with the problem of carbon-fibre wind-turbine waste. Immediately after I graduated, the UK was hit with the 40°C heatwave, and I realised that the problem we were facing was far bigger than I'd been told at school or university. I started reading academic papers on climate science and its social consequences, including mass displacement and social collapse. I realised that my generation was going to face food shortages, extreme weather, and, most likely, a fascist government. This is what caused me to re-evaluate my life, and I decided that I was going to do whatever I could to get as many young people involved in the climate movement as possible.

INVOLVEMENT
Arrested for participating in several actions: blocking Parliament Square in October 2022 by gluing my hand to the road.
Blocking the M25 at J23 for three hours in November 2022 by climbing an overhead gantry (leading to a twenty-four-month prison sentence, of which I served five months).
Leading a march in Liverpool in January 2023 (Edge Lane).
Marching in London in April 2023 (Mile End Road).
Throwing paint and occupying a glass balcony at Exeter University in October 2023.
Marching in London in November 2023 (Holloway and Kennington).
Involved in Just Stop Oil youth mobilisation August 2022 to December 2023, and a founding member of Youth Demand January 2024 to June 2024.

MOTIVATION

I think it just really feels like a responsibility: I can, so I will. I'm really scared about what my life is going to be like as I grow older, and the climate crisis grows worse. By extension, I'm scared about what's going to happen to other people, those I know and those I don't. I don't want to look back at all this shit and think, 'Ah, I wish I'd done something.' As depressing as that sounds, though, I've never felt better about my life. Even when I went to prison, I was content with my decision to take action. Deep down I know that I'm doing the right thing, even if it's really stressful or uncomfortable at times.

INSPIRATION

People like Mahatma Gandhi, Martin Luther King, Nelson Mandela and Emmeline Pankhurst are rightfully celebrated: they dedicated so much to their causes, and there's still so much that we can learn from them.

But I think the people who I'm really inspired by are the ones who didn't 'win' in the conventional sense of getting a demand met. Groups like the ALF (Animal Liberation Front) or the people opposing the Dakota Access Pipeline who have received some of the longest sentences in the history of direct action.

IN NATURE

Every summer, me and some old friends will go to the Isle of Arran or Skye to camp. We regularly see otters, dolphins, eagles, all sorts. No matter what's going on in the rest of my life I'll always feel peace when I'm there.

QUOTATION

'The white conservatives aren't friends of the Negro either, but they at least don't try to hide it. They are like wolves; they show their teeth in a snarl that keeps the Negro always aware of where he stands with them. But the white liberals are foxes, who also show their teeth to the Negro but pretend that they are smiling. The white liberals are more dangerous than the conservatives; they lure the Negro, and as the Negro runs from the growling wolf, he flees into the open jaws of the "smiling" fox.'
Malcolm X

RESOURCES

'Future of the Human Niche' by Chi Xu, Timothy A. Kohler, Timothy M. Lenton, Jens-Christian Svenning and Marten Scheffer. This research paper is available to read: https://www.pnas.org/doi/abs/10.1073/pnas.1910114117.

This was the most shocking piece of research I'd ever read. Maybe it's somewhat inaccessible to those not from a scientific/academic background, but it really paints a picture of how brutal things are going to get.

Alternatively, the film Suffragette (2015), really spoke to me when I watched it. I think in our society we really take social progress for granted, and so many people have forgotten what those women had to go through to bring about votes for women. To me, it makes clear the reality of what we're up against. Campaigning might look different now, but fundamentally we're going to face the same brutality if we want to actually force our government to act with conscience now.

WHAT LIES AHEAD?

Sadly, I think I would have given a very different answer a couple of years ago. I'm not as optimistic as I used to be, but that doesn't mean it's time to stop, quite the opposite. I don't really know, to be honest; I try not to think about it too much. I just hope that when I'm in my 60s there's enough food to eat, and refugee boats aren't being gunned down in the channel. I think I'm only saying that because it's what I'm scared of most to be honest.

6: EVERYTHING'S CONNECTED

'Genocides, ecocides, famines, war, colonialism, rising inequalities and an escalating climate collapse are all interconnected crises that reinforce each other and will lead to unimaginable suffering.'

Greta Thunberg, 11 November 2024

THE REASON I describe the climate crisis as 'the single most significant influence determining the future of our entire species' is simple. It's a view to which every single one of those I worked with on this book subscribes. And it's what makes the challenge of responding – urgently and proportionately – to that existential risk so much more complicated than if climate change was 'just another environmental issue'.

One of the things I most admired about Just Stop Oil was that it never flinched from some deeply controversial conclusions about why we find ourselves in the midst of today's Climate and Nature Emergencies. And why we're never going to be able to respond – urgently and proportionately – to those Emergencies unless we acknowledge the need for a complete transformation of the political and economic systems that have brought us to this moment of extreme peril.

Our discovery, development and scaled-up use of fossil fuels has of course been utterly transformational for humankind. But the economy that pre-dated the first commercial extraction of oil in 1859, as well as the massive expansion in the use of coal since then, already had all the defining hallmarks of the kind of extractive and oppressive economic system that now puts human civilisation at risk.

'When Umbrella was set up in early 2024, I was part of the core team of five that made that decision, and I've been involved since then, representing Youth Demand.

'Our principal focus is on the Gaza campaign, pressing for an embargo on any further arms sales to Israel. This has been by far the most vibrant part of the work we are doing, and the pressure is certainly building on the Foreign Office. But it's hard to imagine this Government doing what it really ought to – morally.

'I feel very comfortable about the comparison made between the genocide in Gaza and the genocide that will happen when the full force of the Climate Emergency destroys the lives of more and more people. I think it's hard for those involved in the Free Palestine movement to think that far into the future, with Israel killing Palestinian men, women and children on a daily basis. But part of the problem is that our movement is incredibly white and incredibly middle-class, and we know we're going to have to move beyond that base to become more effective. We did a small joint campaign during Ramadan, working with Healthcare Workers 4 Palestine, setting up roadblocks during prayers – practical solidarity, if you like! And the police obviously didn't want to go anywhere near anybody praying!'

Sam

'The climate crisis is just one huge symptom of a much bigger crisis: the entire global economy has ended up exclusively benefiting the wealthiest 1 per cent, while everybody else suffers. This is a serious class struggle. So Just Stop Oil couldn't have continued as a single-issue organisation – the crisis today encompasses everything, everything, EVERYTHING!'

Hanan

6: EVERYTHING'S CONNECTED

IGNORING THE ROOT CAUSES

Over the years, I've met countless people in the Green Movement who are reluctant to dig too deep into the root causes of the multiple environmental challenges that we face. We've won a few battles along the way, but we lost the war a long time ago, essentially through our collective inability to address these root causes. Most particularly, our obsessive pursuit of economic growth since the middle of the twentieth century.

There are some uncomfortable realities here: the reality that year-on-year economic growth has been by far the biggest determinant of political success over the last seventy-five years; the reality that conventional economic growth takes little if any account of the damage done to the natural world; and the reality that systematic short-termism means that the ongoing destruction this causes can always be accounted for as 'the unavoidable price to be paid for progress'. These uncomfortable realities have always been there in the background, but never properly accepted as the principal reason why 'environmentalism' has achieved so little over so many decades.

These uncomfortable realities explain why we are now in the midst of the sixth Great Extinction on Planet Earth, with more than one million species at risk, with most critical life-support systems – soil, fresh water, clean air – now stressed to breaking point, and with six of the seven Planetary Boundaries, on which maintaining 'a safe operating space for humankind' depends, already breached.

The truth of it is that it has proved almost impossible to factor this more radical economic analysis into the kind of reactive, case-by-case campaigning that dominates the environment movement, and ends up substituting for any long-term strategy in almost all NGOs.

By contrast, I found no such reluctance amongst young Just Stop Oil activists to acknowledge these root causes. Indeed, they're deeply frustrated that so few climate campaigners are ready to make those connections, let alone to look even further back into the dark history

'It was a significant moment when Umbrella was set up at the start of 2024. The whole idea was to extend the reach of Just Stop Oil, and it's certainly true that there are people who have joined Youth Demand or Assemble who would never have joined Just Stop Oil. Same kind of energy, same enthusiasm for radical change – just a different way of doing things.

'It's encouraging that there is now a much wider recognition of the need for a democratic revolution here in the UK. We need to fundamentally change how democracy "gets done" in this country, in all sorts of different ways, many of which would be underpinned by creating Citizens' Assemblies to start getting more people involved – what is called "deliberative democracy".'

Paul

'In 2022, I was reading a lot more about politics and philosophy, with a growing sense that so many things were just completely "fucked up", with so many different campaigners confronting injustice literally everywhere. That was when I said to myself, "OK, you're out of excuses now, time to do something!" Funnily enough, I signed up for Palestine Action and Just Stop Oil at the same time, met up with a few people from both, and then JSO got back to me to suggest I should do my NVDA training. I then volunteered for a group that was involved in disabling petrol pumps – getting arrested three times in one week.

'So there was no one big moment for me, no "epiphany", more of a slow build as I dug down into the politics, learned about direct action, and then came to the conclusion that it just had to include me. No more of those big official demonstrations that I'd gone on in the past, basically making me feel good about myself, however ineffective they were!'

Phoebe

of colonialism to understand our current crisis. So many of the pathological, massively destructive features of today's global economy can be traced back to that era between 1500, when Spain and Portugal, quickly followed by England, France and the Netherlands, established colonies all around the world, through to the decades of contested decolonisation in the twentieth century. About twelve million Africans were enslaved and transported to the Caribbean and the USA up until the Abolition of the Slave Trade in 1807, with slavery itself not abolished until 1833 in the UK and 1865 in the USA. And throughout that time, inconceivable amounts of natural and mineral wealth were appropriated by those European colonial powers.

So what exactly has any of that got to do with the Climate Emergency, with the sixth Great Extinction and our continuing assault on the natural world? Well, even now, 600 years on, our dominant model of progress is indelibly shaped by pretty much the same extractive, exploitative and abusive norms and behaviours of colonialism, albeit superficially tidied up, legalised, and notionally regulated through today's global economy.

It's never been easy for the NGO community to know how best to adapt to these huge global and ideological forces. As I was finishing the first draft of this chapter in October 2024, WWF International produced its latest 'Living Planet Report' – the first one came out in 1998. The headline finding of 'A System in Peril' was predictably shocking: global wildlife has declined by seventy-three per cent between 1970 and 2020, leaving us 'close to critical tipping points both for Nature and the climate'.

Shocking indeed. Not least because its analysis is so dishonest. As has been the case in every single preceding Living Planet Report, WWF International cannot bring itself to tell people the truth: that it suits all world leaders, democrats and autocrats alike, to continue to fuel this engine of destruction in the name of year-on-year economic growth; that the system depends on maintaining abhorrent levels of inequality, and that the super-rich have been allowed to become a law unto themselves; that there are just too many people, mostly in the rich world, consuming too many resources and fossil fuels on a finite and flammable planet;

'We are very aware that when we succeed in ending the use of fossil fuels, that's not going to be the end of it. Our dependence on fossil fuels is just a symptom of a broken political system, and if we don't deal with that system at a deeper level, then we'll just go from putting out one fire to putting out another. For a lot of people, the climate crisis just isn't their top priority – which is hardly surprising when it's consistently framed as something out there in the future – because there is so much else going on in their lives. That's why it's important that we think about the connections within a deeper framework, whether that's genocide in Gaza, wealth inequality or our broken democracy.'

Olive

'Assemble is an important part of Umbrella, advocating for Citizens' Assemblies and replacing the House of Lords with a House of Citizens. And we still need some pretty radical campaigning on proportional representation! That's where I'm most involved at the moment, trying to persuade a lot of people that our democracy is so screwed that it can't possibly deliver what people want in terms of fairness and proper representation of people's views, let alone dealing with poverty, the environment, the climate crisis and so on. There's so much work to be done getting people to understand the importance of more direct, more participative forms of democracy.

'The campaign around the ongoing genocide in Gaza has really made me think. The outpouring of solidarity and support for the people of Gaza has been deeply moving and has helped a lot of people to see the connection between that genocide and the genocide that will now inevitably occur in the future through climate breakdown. But it's also been difficult to think through the implications of that comparison – how do we get people to connect emotionally with the fact that the climate genocide will be hundreds of times worse than what is happening in Gaza? I just don't think we've been able to achieve that.'

Eddie

and that life on Earth is dying as a direct result of this political insanity.

Even to hint at such inconvenient political realities would be far too contentious for WWF, putting out of joint the noses of far too many influential people on whom its continuing support and fundraising depend. Like every other major international environment or conservation NGO in the world, trotting out the same banal and dishonest half-truths, WWF International can reasonably be accused of total complicity in the resulting, seemingly inexorable slide into ecological breakdown.

As revealed in their interviews with me, young Just Stop Oil activists have a pretty jaundiced view of mainstream environmentalism and conventional climate campaigning. And pretty much everything that happened in the second half of 2024 seemed to confirm that, leaving them with even less reason to assume that the Moderate Flank of the movement has anything much to offer young people today.

The year ended with five crushing setbacks: the big biodiversity conference in Colombia in October ended in failure, with zero progress made on the '30 by 30' (thirty per cent of lands and oceans to be fully protected by 2030) commitment made the year before; the even bigger climate conference in Azerbaijan in November ended in failure, with rich-world nations doggedly refusing to commit to the level of funding now required to protect poorer nations from the ravages of a climate crisis to which they have contributed little; 2024 was confirmed as the hottest year ever, with an average temperature of about 1.6°C, making a mockery of the stubborn optimists' belief in 'keeping 1.5°C alive'; negotiations around the Global Plastic Treaty in December ended in failure, destroyed by the refusal of oil companies to put a mandatory cap on production; and Donald Trump was elected as the forty-seventh President of the USA.

Electorally, things didn't work out well in 2024 elsewhere. The continuing rise of populist right-wing parties in Europe had a big impact on the EU Parliamentary elections in June. There were already clear indications of pushback on a host of climate and biodiversity initiatives

'I think quite a lot of us are "maxed out" on keeping the focus purely on the fossil-fuel industry ... our fixation on fossil fuels is just a symptom of so many other things. Involvement in Just Stop Oil has allowed people to see the massive social justice issues involved, not just the environment and climate; the oil and gas industry is just the filthiest aspect of a filthy and corrupting system, which systematically deprives people of the right to be involved in the decisions that affect them most.

So I'm quite involved now in the Assemble initiative, engaging with citizens where they live – "meeting people where they're at". Sounds so simple, but it's much harder than people realise – and that isn't just because we're doing some of that citizen engagement in Tunbridge Wells!

So much of what doesn't work these days can be linked to the patriarchy. So many powerful men with crazy control fantasies and insufferable egos. And what is it about those billionaires who fantasise endlessly about the prospects for human beings ending up on a lifeless rock somewhere in the distant solar system, rather than joyfully embracing the Earth in all its wondrous diversity and beauty? They seem to have so little compassion for their fellow citizens, so little love for the Earth and all the non-human creatures that we share it with.'

Daniel H.

'We have to do a better job bringing people with us – I'm not sure how much all the talk about "revolution" necessarily helps! Inevitably, a lot of what they hear from us represents "the end of the world" as they know it. We have to do more about giving voice to the "better world" that is such an important part of what we're trying to achieve. That would help make a much stronger connection with our NVDA strategy. The emphasis on NVDA doesn't actually create the tension, but it certainly brings the tensions that are already there to the surface, forcing people to see things in a different way. But it doesn't necessarily point to a longer-term resolution – or even solutions.'

Avery Simard

by the end of the year, with centre and centre-right politicians only too keen to make concessions to keep the far right at bay. Climate change has slipped down the agenda in many countries, with waning public interest, despite quite shocking climate-induced disasters demonstrating time after time that this has to be the time to step up not step back.

MAKING THE CONNECTIONS

It's just got a lot harder – for everyone. But even before then, Just Stop Oil was struggling to broaden its supporter base, and was criticised by some on the Left for its reluctance to identify explicitly with wider political struggles. By the end of 2023, there was a lot of pressure internally suggesting that Just Stop Oil needed to move on from its original and exclusive demand of 'no new oil and gas licences in the North Sea' and to take on a range of equally pressing social and political issues. In January 2024, the decision was taken to create Umbrella, made up of four separate elements: Just Stop Oil itself; Youth Demand; Assemble; and Robin Hood, as a campaign to target the super-rich – which was subsequently dropped for lack of resources.

YOUTH DEMAND: COMBATING GENOCIDE

The principal focus of Youth Demand (formerly Just Stop Oil Youth) has been to join forces with Palestinian NGOs addressing the ongoing genocide in Gaza.

On 7 October 2023, Hamas carried out its horrendous assault on Israel. 1,139 people were killed, and 695 taken as hostages, including thirty-eight children. Deepest revulsion was the only possible response, accompanied by a recognition that Israel had the right to defend itself and to seek to go after Hamas in such a way that a horror story of this kind could never happen again. In pursuit of that goal, Israel occupied

'Everything is completely interconnected, with so many issues coming from the same roots: capitalism, colonialism, extractivism and the exploitation of people and the land. When you're fighting for one of these, you're fighting for all of them. People in the climate movement see that, but people outside often don't see it at all. Even with something like Gaza and Palestine. With the climate crisis, we talk all the time about the impact it will have on people in the future, and the suffering it will cause, how many people are going to end up in drought and famine and so on. But that's already happening in Palestine, right now, having their homes destroyed, with everything they know and love collapsing around them.

'To me, these issues are inseparably linked. And we have to show that with our actions, not just with our words.'

Eilidh

'From a very young age I was constantly being told that I had "a strong sense of justice", and a corresponding dislike of authority! I dropped out of school before A levels, living somewhat precariously with a group of young friends. I was referred by Social Services to Youth Support, and my youth worker basically saved my life, helping me get into the Academy of Contemporary Music, and from there to Brighton University.

'I was involved in a big campaign at uni to try and stop 110 redundancies going ahead, and found this really empowering. I was then arrested for blockading an arms factory involved in sending arms to Israel. But I gradually got more involved in Just Stop Oil, working with them on helping with the legal campaigns in several actions, in other support roles and so on.'

Cole

6: EVERYTHING'S CONNECTED

the Gaza Strip a few weeks later, declaring that nothing less than the total elimination of Hamas would be sufficient. Hell on Earth was subsequently unleashed.

Day after bloody day, the world watched on as 'a genocide in real time' was carried out by Israel, with the sought-after destruction of Hamas 'providing cover' for Israel, the USA and the UK to enable and prolong that genocide, politically at the UN and financially through the sale or arms, energy and other critical supplies.

In early 2024, Youth Demand added its voice to the campaign for a two-way embargo on arms sales to and from Israel. This demand was contemptuously rejected by the Tory Government. When Labour was elected in July 2024, it initiated a review of arms licences to Israel and, in September 2024, thirty licences were suspended. By virtue of the fact that around 330 licences are still in place, including licences for critical components of the F-35 fighters that have caused such devastation in Gaza since the start of the war, the Labour Government is seen by Youth Demand as no less complicit in the Gaza genocide than the Conservative Government.

Throughout 2024, Youth Demand stood shoulder to shoulder with Palestine Action, Free Palestine and other organisations. They supported them on marches, in actions and through joint communications.

But has this been a two-way street? Controversially, Youth Demand and Just Stop Oil set out to make a comparison with the genocide in Gaza and the now inevitable death of hundreds of millions of people in the future as a consequence of extreme climate disruption, as powerfully confirmed by The Institute and Faculty of Actuaries (see p. 43). They regularly accused the Government of being 'guilty of mass murder' as a consequence of its continuing support for fossil fuels, deliberately going out of their way to cause maximum offence, amidst accusations by Ministers and the media of 'grotesque and melodramatic exaggeration'.

What Youth Demand is referring to here is what is known as 'murder by oblique intent', as spelt out in Article 30 of the Rome Statute of the International Criminal Court. This states that death or harm doesn't have to be a primary intention for someone later to be held criminally responsible: they simply have to be 'aware that it will occur in the ordinary course of events'. Here's how Youth Demand interprets that Article:

> By issuing licences for new oil and gas, politicians know that they will breach agreed climate targets, contributing to millions and ultimately billions of deaths. This unfathomable death toll, present and future, is known to politicians as a 'consequence that will occur in the ordinary course of events'. It is a law of nature that burning fossil fuels after a certain point kills people – and burning more kills more. When whole peoples are erased from the surface of the Earth, as will happen when Island States are obliterated, it is genocide.

The logic of 'oblique intent' is there. It would obviously have to be tested in law, as Just Stop Oil sought to do back in 2023, when it urged the Metropolitan Police to address the 'real criminals' in the fossil-fuel companies rather than climate protesters. The UK Youth Climate Coalition is seeking to test this provision at the International Climate Court in the Hague, presenting a compelling dossier of evidence against BP.

Youth Demand acknowledges that it's not been plain sailing persuading their colleagues in Palestine Action to see complete equivalence between these two genocides. In a rather uncomfortable way, this can be seen as yet another manifestation of the 'tragedy of the horizon' (see p. 69), as it's not easy for anyone to compare the shocking suffering of millions of people in Gaza – deadly, remorseless and happening right now – with the putative death of hundreds of millions of faceless, nameless but equally blameless people as climate breakdown worsens in the future. Inevitable though that now is.

Israel's war against the people of Gaza has also provided clear evidence of ecocide – defined as 'unlawful or wanton acts committed with knowledge that there is a substantial likelihood of severe and either widespread or long-term damage to the environment being caused by these acts'. Inconceivable numbers of bombs and shells have been dropped on or fired in Gaza; 70 per cent of Gaza's buildings have been badly damaged or destroyed; 80 per cent of agricultural land has been badly damaged, and those parts of its water system still intact have been chronically contaminated. Wastewater treatment plants have been forced to shut down due to lack of power – Israel controls more than 90 per cent of Gaza's energy – with untreated sewage impacting horrendously on the lives of most of Gaza's citizens.

Since 2010, a determined group of campaigners, Stop Ecocide International, has made astonishing progress in getting ecocide accepted as a crime, starting with that now widely accepted definition of ecocide and initiating the process through the International Criminal Court to have the crime included within the Rome Statute alongside the four crimes currently listed: genocide; crimes against humanity; war crimes; and the crime of aggression. A clearer and more abhorrent demonstration of what ecocide means, inflicted remorselessly on the people of Gaza and the land they love, is impossible to imagine.

Our world has been systematically ravaged by the activities of the fossil-fuel incumbency over the last 150 years, which has benefited economically to an extraordinary degree during that time. The massive new investments in fossil fuels that are still being taken forward, through both Independent and National Capital Oil Companies, guarantee accelerating climate change and ecological breakdown in the future, which in turn all but guarantees widespread social disruption and potential societal collapse. This poses a far graver threat to the integrity and resilience of our democracies than anything experienced in modern times.

This has been a recurring theme in many of the interviews conducted for this book, revealing an acute awareness of the critical importance of democracy in ensuring that we avoid any descent into authoritarianism as a consequence of extreme climate change, and of the imperative need to defend democracy in the event of such a breakdown.

'So many people today are distracted by all the marketing with which they're being bombarded, every minute of the day, with the endless false promises that their lives will be just a little bit better if they buy this or that. It cuts people off from reality.

'And it certainly stops people making the connections between all the different challenges that we face. One of the reasons I was first drawn to Just Stop Oil was the very clear focus on one campaign goal – no new oil and gas licences. But the climate crisis isn't a single-issue campaign; it's linked to so many other aspects of our lives, economically, politically, historically. I started to see things in a much more systemic way when I first got involved in climate activism, and that's both empowering and exhausting!'

Emma

'There are a lot of people involved in climate campaigning who say we should keep a really tight focus on getting rid of fossil fuels – just stick to that! But I think this is both nonsensical and illogical. The principal reason we're facing this crisis is the super-rich and the political class, who, for the sake of their status, power and wealth, seem to be quite comfortable with the idea of killing a lot of innocent people in the future. It's not just the countless areas which are going to be directly affected by heatwaves, floods and droughts, but people having to leave their homes – as climate refugees. And then you have to think about all the social conflict that will result from these climate impacts, with a lot of people ending up being killed, not because they died of heat stroke.

'Which is why we really have to think about Gaza. If you're concerned about the climate crisis, you should be absolutely horrified about the situation in Gaza, which is just telling us what it's going to be like in the future. I think there has to be some kind of moral deficit in those people who are very concerned about the climate crisis but not about Gaza.'

Daniel K.

6: EVERYTHING'S CONNECTED

ASSEMBLE: DEFENDING DEMOCRACY

One of XR's original demands back in 2018 was to establish a Citizens' Assembly that would determine more appropriate ways of addressing the Climate Emergency. In January 2020, partly in response to those demands, the House of Commons, through six of its Select Committees, initiated the UK Climate Assembly. This is widely recognised to have been an exemplary process, with a group of 106 randomly selected UK citizens engaging with independent experts across a very wide range of climate issues. Unfortunately, its excellent final report in September 2020, 'The Path to Net Zero', was totally ignored by Ministers and by the Government as a whole.

That kind of contemptuous disregard for legitimate governance processes of this kind, and the ease with which the Government simply ignores the strength of feeling that people have regarding the climate crisis, became a significant element in Just Stop Oil's theory of change – and a critical justification for the use of disruptive NVDA. Ironically, critics of NVDA accuse those involved of being 'anti-democratic', using extra-parliamentary tactics 'to try and force their extreme and unrepresentative views on democratically elected politicians'. As we'll see in the next chapter, this debate is as old as the hills, but it was a core component of Just Stop Oil's advocacy that the Climate Emergency will never be properly addressed until our democratic institutions and practices can be dramatically reinvigorated.

Assemble was set up at the same time as Youth Demand in January 2024. Its principal demand is that the House of Lords should be replaced with a 'House of the People', as a standing Citizens' Assembly, and that local citizens' assemblies should be established wherever possible to stimulate informed debate and appropriate action plans on the ground.

That all makes good sense, but all this is playing out against a backdrop where more and more young people seem to be losing faith in democracy altogether, with all sorts of polls and surveys showing that 'disaffection' can all too easily lead to a sense of disenfranchisement. As

ANNA HOLLAND
Poems from Prison

Floodplain

And who will think of the prisoners
when the walls, built to keep us in,
cannot keep the water out?

When floods seep through vents
will someone set us free?

This prison was built on a floodplain.

So the fear of heavy rain
is a punishment that comes free:
at no extra cost
   to the taxpayer's money.

With a clap of thunder we feel it –
the airless silence
of a hundred inhales.

A flash of lightning illuminates
   our faces
pressed against bars
   built into windows.

We count the puddles,
   guess their width
wonder if locks turn when
underwater.

I do not know these women,
but I have heard their prayers.

There Is No Art

There is no art
so beautiful
as soup on a frame.

No portrait stands so tall
as a student's first steps
on the asphalt road.

There is no song
so melodic
as women chanting in the streets.

The scientist's call
is more rhythmic
than any ode or sonnet.

When the revolution finds its canvas
you'll see a self-portrait
is demanded.

There is no art
so beautiful
as action.

many as 20 per cent of young people are sympathetic to the idea that what we need here in the UK today is 'a strong leader who doesn't have to bother with elections'.

Nothing could be further from the views of those I've worked with on this book. A number of those I interviewed clearly long for a deeper more 'revolutionary' process, often talking quite wistfully of the extraordinary but little known story of Rojava (the Kurdish region of North and East Syria) which has successfully sustained a form of libertarian socialism based on gender equality and decentralization since 2010, in often unbelievably difficult circumstances. This provides a unique model of how different things might be, but that region is now in a state of even greater turmoil since the fall of the Assad regime in December 2024.

But it's hardly surprising to them that so many young people feel deep anger after more than fifteen years of austerity: declining public services, particularly youth services and mental health care; a lack of secure jobs; escalating rents and very poor quality housing; punishing debts for graduates; real cost-of-living impacts, especially with regard to food and public transport; and child poverty levels at an all-time high. And they see the rich continuing to get richer even as Labour imposes another round of austerity – and wonder exactly who the whole system is designed for.

In May 2013, Peter Mandelson, now the UK's Ambassador in Washington, declared that New Labour was 'intensely relaxed about people getting filthy rich as long as they pay their taxes'.

In 2025, Keir Starmer and Rachel Reeves appear to be equally relaxed. In 2024, the number of billionaires in the UK increased by a further four, making a total of fifty-seven – a mere two per cent of the global total of 2,769 billionaires – as revealed in Oxfam's 'Takers Not Makers' report in January 2025.

It was part of the original planning for Umbrella in 2024 that it would join forces with others here in the UK to make the case for wealth taxes, part of the revenues from which would be invested directly in an accelerated exit from fossil fuels. It was subsequently decided not to proceed with this campaign – we're still talking about tiny numbers of people in Umbrella, with very limited resources, taking on massive societal and economic challenges.

As a member of the Green Party for the past fifty years, the only mainstream party that challenges the total dominance of economic growth as a measure of progress, and the author of *Capitalism as if the World Matters*, a somewhat 'have your cake and eat it' book about capitalism, I know only too well how hard it is to challenge the deeper norms and ideological constructs which underpin today's neoliberal capitalism.

This is not just on Just Stop Oil. The combined fire-power of the entire Green Movement, and all those involved in climate campaigning over the past thirty years, has had next to zero impact on forcing politicians to think so much more deeply about radical economic transformation.

HARRISON DONNELLY

I'm a philosophy student. I love chess, I love partying, I enjoy biking, I enjoy music, I love my family and my friends. I was presented with the harsh reality of the climate crisis in primary school. Before I took action in London through The Rokeby Venus action, I asked, 'How many more people need to die until we take action?'

INVOLVEMENT

I first joined Just Stop Oil in 2023 where I attended a student meet-up. I marched on the roads firstly, and the next day I was wrongfully arrested outside the TotalEnergies headquarters in Canary Wharf.

In October 2023, I took part in an action at Birmingham University where I held a rally and threw paint over the library windows. I was subsequently arrested and went to court, where I was ordered to pay a fine.

In November 2023, I took part in The Rokeby Venus action in London (with Hanan Ameur – see pp. 163 & 164), where we peacefully smashed the protective glass covering the painting. We were subsequently arrested and are currently awaiting trial.

MOTIVATION

I think we're at a crucial tipping point in the survival of life on earth. We can either sit back, relax, continue on with business as usual and allow the world to burn, or we can realise it's not too late, that we can still limit this. How many more people need to die?

What motivates me is knowing I want to have kids in the future, but that it would be immoral to bring life into this world whilst knowing I may not be able to protect them when inevitable floods, food shortages, droughts and wildfires come for them. What do I tell my future kids when they ask me at night, 'What about the flood that's coming?' I need to be able to tell them that floods only happen in bad dreams and not in the real world.

INSPIRATION
If I had to choose, I would simply say that Daniel Knorr, a current JSO activist, inspired me the most. He has put his liberty on the line time and time again. He is a wonderful young man, and his passion for resisting against the climate crisis is contagious. As I write this, he is still wrongfully imprisoned.

IN NATURE
There is a small reservoir next to where I live and I enjoy taking walks there every day when I am away from university. There are no roads, no buildings, just green fields as far as the eye can see. I feel at home there. I am a strong believer in the idea that nature can heal wounds, because we are all intrinsically a part of nature. To destroy nature and/or the natural world, you are destroying the very origin of your own self.

QUOTATION
'It's our future. We'll take it from here.' Arthur, JSO activist

RESOURCES
How to Blow Up a Pipeline, By Andreas Malm.

WHAT LIES AHEAD?
I'm very optimistic. I believe that one day we will have a healthcare system that operates to keep people healthy, and people will care for one another as they'd like to be cared for. We'll have green grass everywhere, floating trains, renewable energy – the classic utopian idea. I hope that one day every single person will act out of compassion for one another, because we all feel the same horror and disgust as we see people that we don't know being washed away by floods. These people are mothers, brothers, uncles, friends, families, partners – I dream of a world where floods, and other disasters, only happen in bad dreams, not where floods are caused by businessmen flying 48,000 feet above the floods and wildfires.

INDIGO RUMBELOW

I was brought up on the Gower Peninsula in Wales. We kept chickens, geese and an array of other animals, and also grew veg. I was interested in politics and movements early on. I was outraged by the treatment of Bradley Manning and Julian Assange. The first protest I went on was the Student Fees demonstrations in 2012. When I moved to Brighton to study art and design, I became interested in the movement against fracking, but I was very much into my student life at that time and still didn't really understand how badly our government was failing us.

INVOLVEMENT

In 2017, I visited Germany to attend an action organised by Ende Gelände and Arid Action against opencast coal mining. We were pepper-sprayed by the police, and then blocked a train line before being arrested. There were so many of us we were all released without charge. I was involved in a similar campaign in Pont Valley, County Durham, against a vast new opencast coal mine.

I then got involved in the anti-fracking movement in the UK. Following hundreds of demonstrations, in 2018, the Government issued a moratorium on fracking, showing that NVDA really can work.

I got involved in Extinction Rebellion actions pretty much from the start, later training as a Legal Observer, then an Actions and later an Arts Co-ordinator. I was arrested in 2019, at an advertising festival, and later at London City airport.

Having helped set up Insulate Britain, I was part of the M25 blockades and was arrested a number of times and fined. With Just Stop Oil, I've been arrested multiple times, mainly on conspiracy charges, and have also been involved in a number of different actions, including singing carols outside the Prime Minister's home, for which we were acquitted!

I am currently in HMP Styal (having been found guilty of conspiracy to cause a public nuisance planning to disrupt flights at Manchester Airport), serving a sentence of two and a half years.

MOTIVATION

It's very simple: the only thing to do in a system that is driving itself to extinction is to rebel. And if we look back at what history has to teach us, it's clear that the only way of creating the kind of urgent, non-linear change we need is for people to commit civil disobedience to force that change through. That's what has driven me since I first got involved in Extinction Rebellion.

INSPIRATION

One of the most inspiring people I know is Angie Zelter, who's been involved in the peace movement for decades. She founded the Snowball Campaign in the 1980s, cutting wire fences around US military bases. And then she went on to smash up a BAE Hawk jet that was about to be exported to Indonesia, and with a group called Trident Ploughshares caused significant damage to a key nuclear facility. She's a huge supporter of the Palestinian cause and was also one of the first people arrested in XR's protests in London in 2019. She's amazing! Her book is called Activism for Life.

IN NATURE

During my childhood in Wales, near the Loughor Estuary, I spent so many happy days knee-deep in the mud, finding sheep skulls and so on! Sometimes, when I'm surrounded by four grey walls in prison I take myself off on imaginary walks around the estuary.

RESOURCES
There's so much I could mention here! Starting with Rich Felgate's film Finite: The Climate of Change. It amazes me how few people know about all the brilliant campaigning work that went on here in the UK to put an end to opencast coal mining!

Then there's The Shock of the Anthropocene, by Christophe Bonneuil and Jean-Baptiste Fressoz, and an extraordinary book, You Have Not Yet Been Defeated by a British/Egyptian political activist and blogger, Alaa Abd El-Fattah, who's been in and out of prison since 2006.

WHAT LIES AHEAD?
It's not easy trying to imagine where we'll be in a couple of years' time, especially as we're still going to be seeing JSO activists prosecuted all the way through into 2026 and possibly 2027 – just because of congestion in the criminal-justice system. And it's possible that the whole climate movement might evolve into something different, fighting on a different battle-front, connecting more closely with those concerned about the massive social injustice in this country, linking with people who are more vulnerable to climate disaster even though they've had nothing to do with causing it. Whatever happens, civil disobedience will be as important as ever.

(See pp. 26 & 27 for Indigo's account of the start of Just Stop Oil.)

'We all draw a lot of strength from those who have gone before us. I was on remand in prison just before the general election in July 2024, so was deprived of my vote! As it happens, I'd borrowed a book called *March, Women, March*, by Lucinda Hawksley, which includes a brilliant account of the Suffragettes. The irony was not lost on me!'

Cole

'So many of the future effects of the climate crisis are now completely locked in. This terrifies me. It's difficult to maintain a positive vision of the future, even though I still hang on to the belief that change can come monumentally fast, in the right circumstances. But climate breakdown is different from all those other massively important progressive causes – for instance, even the Suffragettes always recognised that their campaign was not time-limited. If it hadn't been for the First World War, they would have gone on campaigning until they won. With the climate, we have a lot of deadlines built in as regards various climate tipping points, which is what makes it so frustrating when the politicians just go on and on wasting so much time.'

Alex

'It's not easy showing all the links through Umbrella. The campaigns are pretty separate. With Assemble we have to be able to show what a properly functioning democracy would look like, and how we could really build communities differently. Even with Youth Demand and the shared campaign with Palestine Action, it's been really difficult for us to make the links back to Just Stop Oil's core demand: getting rid of fossil fuels. But I really think we have to persevere with this in terms of the actions we take.'

Avery

7: TAKING DIRECT ACTION

'Activism is the rent I pay for living on this planet.'
Alice Walker, 1989

THE CLIMATE CRISIS is not something that can be read about, reflected on and then set aside. Each of us, individually, with our hearts and consciences fully deployed as much as our intellects, have to keep checking in on our own level of commitment at any particular moment. The extracts from some of the interviews I did for this chapter speak very powerfully to those moments of personal awareness.

There often comes a time in any environment or climate campaign when the mainstream consensus as to what needs to be done can be seen, by pretty much all and sundry, to be failing. A few climate scientists, some campaigners and quite a few hard-core eco-doomsters came to the conclusion that failure was inevitable decades ago. For others, myself included, it has always been a question of the gap between what the science tells us and what the politics tells us – and whether that gap is narrowing or widening.

When the Tories (and the deluded LibDems, as they were back then) got elected in 2010, that got me seriously worried. Up until then, I had thought that the gap was just about bridgeable. Labour had done pretty well in office. Then we had the much-hyped 2015 Paris Agreement, by which I was largely unpersuaded, followed by the Brexit referendum in 2016, followed by the election of Donald Trump. At that point it was clear to me that the combined campaigning efforts of NGOs, scientists, progressive businesses and international diplomatic processes had pretty much run out of steam. Our go-to 'theory of change' was looking more and more threadbare, bordering on demonstrably fucked.

'I was in some kind of "soft denial" until the end of 2021, thinking that "someone's got this". I don't know which rock I was living under at the time, but it was a big one! Then the 40°C heatwave in the UK really hit me in the face, and I realised that "nobody has got this". At Oxford I got a bit involved with Extinction Rebellion, leafletting, marches, that kind of thing – before I began to follow much more closely what was happening with Just Stop Oil at the end of 2022. They were blocking streets in London at the time, and there was an extraordinary film circulating amongst campaigners of a Range Rover almost driving over a Just Stop Oil protester. And then another woman went to sit right next to her. I was very taken by that kind of emotional courage. So after my non-violence training, I did my first action with them in January 2023.'

Daniel K.

'There's something special about the way we work together, which allows us not to get too bothered about the often hostile reactions of mainstream environmentalists, let alone the right-wing media! We just have to put up with that and I often recall that Extinction Rebellion slogan: "We are not here to be liked." We can't waste time trying to persuade those environmentalists who don't want to know – in due course they will embrace us!

'Being together on these actions has been a remarkable experience. I'm not a spiritual person, but I found real spiritual depth in those actions, in the closeness of the relationships, in the way courage calls to courage. That's one of the reasons why I'm not too worried about whether or not Just Stop Oil continues as an organisation – it doesn't really matter. If it disappears, it will be just like a snake shedding its skin. The ideas, values and absolute commitment to non-violence will all still be there.'

Daniel H.

7: TAKING DIRECT ACTION

So when XR erupted in our midst in 2018, I was more than relieved. I joyfully observed some of the London actions in person, though not as an 'arrestable' participant. I was still full-time with Forum for the Future at that stage. Then came Covid. XR all but collapsed, and I profoundly disagreed with the tactics adopted by its successor, Insulate Britain. It took me a while to find an authentic way of supporting Just Stop Oil. And now it too has gone.

THE STATUS OF NON-VIOLENT DIRECT ACTION

Let's wind back a bit here. As I suggested in Chapter 1, all protest movements operate across a tactical spectrum, with an Establishment Base, working away at changing policy from inside the system; a Moderate Flank, working from outside the system, but constructively, always with a view to establishing common ground; a bunch of shouty, in-your-face campaigners, who absolutely get it, but have little interest in helping others to get it; and then, sometimes, a Radical Flank made up of people and organisations choosing to take Non-Violent Direct Action (NVDA) to accelerate the process of change. In my world, Greenpeace filled that radical space for thirty years or more. I loved the actions they took – especially when I was Director of Friends of the Earth, which was always much more nervous about using NVDA!

The campaigning scene has changed dramatically since then. And what XR and Just Stop Oil have done is part of a huge patchwork of 'climate rebels' all over the world, loosely affiliated to two different international networks. At the last count, the A22 network includes organisations taking NVDA in around a dozen countries, including Finland, Sweden, Norway, Germany, Austria, the Netherlands and … Uganda.

Students Against the East Africa Crude Oil Pipeline (EACOP) is one of the most remarkable groups in the network, with dozens of its members regularly, and violently, arrested for their involvement in peaceful

'It made sense for me to get involved in Just Stop Oil in 2022, when I realised change wasn't likely to come without that kind of direct action. In March, I was involved in the slow marches for the best part of a month, when I felt I could align my understanding of the climate crisis with my moral concerns and the action I was taking. I really did find a sense of inner peace, despite being in the most chaotic circumstances, blocking a road with drivers screaming at you, the police being brutal, and knowing that you'd be doing it again the next day. It was incredibly intense.

'I was a spokesperson for Just Stop Oil at that time, but felt increasingly uncomfortable being singled out as an individual. You get held to an almost impossible standard, with critics trying to pick away at any perceived hypocrisy on your part, as well as facing extraordinary hostility on Twitter and social media. There were horrible things being said all the time, death threats and so on.

'In October that year, I was diagnosed with Type 1 diabetes. There's no doubt that all the stress I'd been feeling contributed to this. As you know, lots of climate campaigners end up burning out, finding themselves in a place where there are no checks and balances about being an activist. It's not like a job where you get proper training!'

Emma

'My loyalty is indeed to Just Stop Oil, simply because that's where I first got involved and began to understand the real importance of civil disobedience. But it's the *ideas* that really matter, and the heritage we benefit from through so many social movements in history. Personally, I feel a strong affinity with the Civil Rights Movement and Martin Luther King. But there are so many non-violent revolutionary struggles that inspire me – the overthrow of Milošević in Yugoslavia, the Arab Spring, and so on.'

Sam

marches, delivering petitions and picketing government buildings. The level of brutality, both at the point of arrest and in prison, is startling. But the students' determination to bring an end to the insanely destructive EACOP project, which would displace over 100,000 people, destroy several nature reserves and – if it goes ahead – cause around 35 million tonnes of CO_2 to be emitted every year, is remarkable. If you don't know about their campaign, please do take a look: https://www.eacop.com/.

Victories or partial breakthroughs are hard won – and correspondingly cherished. In the Netherlands, XR NL started campaigning against the Dutch Government's continuing subsidies for its oil and gas industry back in June 2020, gaining a lot of public attention over the next couple of years. But nothing changed – apart from the revelation that the scale of these subsidies, at $50 billion a year, was ten times greater than initially thought!

So it began blockading a section of the A12 motorway near The Hague in July 2022, with subsequent actions attracting more and more people – and more and more arrests. There were 1,500 arrests in May 2023, and the campaign carried on until October that year, when an important symbolic victory was achieved as the Dutch Parliament passed a motion instructing the Minister of Climate Policy and Green Growth to prepare a timetable for phasing out all subsidies. Not a total victory, but substantive enough to be able to put the protests on hold until March 2025, when campaign organisers decided that insufficient progress had been made in eliminating those subsidies.

Different organisations draw their inspiration from many different sources, past and present. Some of the most interesting conversations I've had with Just Stop Oil colleagues have been in that context: given the extraordinary role that civil disobedience has played in so many different social movements, in so many different countries, which matters most to you? It was a close-run thing between the Suffragettes in the UK and the Civil Rights Movement in the US! But the Moderate Flank/Radical Flank analogy, as explained in Chapter 1, made the Suffragettes particularly compelling for many Just Stop Oil activists.

'I've learnt so much reading about the Suffragettes and the Civil Rights Movement in the USA. Every single one of the rights available to me today as a queer woman has been fought for through civil disobedience, through ordinary people putting their bodies on the line, in ordinary places, doing extraordinary things as a community.

'But for me, it's still difficult making comparisons between movements. The Climate Emergency is different, because of the time pressure involved – I might not be comfortable using that phrase "too late", but it's already too late for some things for millions of people, and it will soon be much too late for many more millions who will lose their lives in any climate breakdown. That makes it different for us.

'I do look up to an organisation called Otpor!, the Serbian student movement (see p. 171). The way they took on the Milošević regime is really inspirational!'

Ella

'I've learned a lot about resilience and the ability to deal with many different situations. I started to appreciate just how much non-violence actually changes the way I interact with people around me, and a lot of that comes from the experience of slow marching. When you're face to face with someone getting increasingly aggressive, shouting out that you should die because you're blocking their car, you certainly have to think differently. It does get easier not to feel afraid in those situations, to trust yourself and to feel compassion for all those caught up in this crisis, in one way or another.'

Olive

7: TAKING DIRECT ACTION

THE SUFFRAGETTES

These conversations made me go back and revisit that astonishing decade – between 1903, when the Women's Social and Political Union (WSPU) was set up by Emmeline Pankhurst and a handful of colleagues, and 1914, when the onset of the First World War caused the WSPU to 'cease all hostilities' with the Government. By then, the Suffragettes were anything but a non-violent movement. Indeed, some historians believe they should more accurately be described as a terrorist organisation.

When the WSPU first broke away from the mainstream suffragist movement, which had spent the best part of forty years petitioning for votes for women, but had by then in effect given up on the possibility of the Government ever introducing a Suffrage Bill, it still subscribed to the movement's strict non-violence principles. But as support for the WSPU grew – up to 500,000 women gathered in Hyde Park in 1907 to press their demands – the Government used every devious, dishonest means to string them along, leading to more and more women being prepared to be arrested for marching, disregarding police edicts and interrupting political meetings.

Throughout this time, there was a majority of MPs in the House of Commons who were personally pledged to vote in favour of a Suffrage Bill, as well as a majority of voters in the country. But this was felt by the WSPU to be 'no longer of the slightest use, however sincerely felt', simply because the Government remained absolutely intransigent – accusing the Suffragettes of being 'self-made martyrs', to which Emmeline Pankhurst replied:

> We never went to prison to become martyrs. We went there in order that we might obtain the rights of citizenship. We were willing to break the law so that we might force men to give us the right to make law.

The 'climate of violence' at that time is extraordinary to reflect on. Horrendous violence, including widespread sexual assault, was used by the police during protests in 1910, particularly on what became known as Black Friday. From that point on, especially after the Government escalated levels of violence by resorting to 'forcible feeding' in an effort to stop the hunger strikes by imprisoned Suffragettes, all acts of violence were described by the Suffragettes as 'acts of war', not as crimes, with imprisoned Suffragettes described as 'prisoners of war' rather than as common criminals. In their defence, the comparison was always made with the violence used by men in their own suffrage struggles, reminding people that, in Bristol alone, men seeking the franchise had 'burnt down the Mansion House, the Custom House, the Bishop's Palace, the Excise Office, three prisons and much private property', suggesting that the Government 'may indeed be thankful for the moderation shown by women'.

The year 1913 was one of 'unrelenting violence'. In February, Lloyd George's newly constructed house was destroyed by a bomb, and Emmeline Pankhurst was found guilty of 'counselling and procuring the perpetrator', receiving a three-year sentence. In June, Emily Wilding Davison died after being trampled underfoot by the King's horse at the Epsom Derby. Arrested Suffragettes started 'thirst strikes' as well as hunger strikes. In March 1914, Mary Richardson used a meat cleaver to slash the famous *The Rokeby Venus* by Velázquez in the National Gallery, declaring in court:

> I care more for justice than I do for art, and I firmly believe that when a nation shuts its eyes to injustice and prefers to have women who are fighting for justice ill-treated and tortured, that such action as mine should be understandable. I don't say excusable, but it should be understood.

This was, of course, the inspiration for the action taken by two Just Stop Oil activists in November 2023, attacking the protective glass in front of the same painting.
(See testimonies from Hanan Ameur and Harrison Donnelly on the following pages.)

HANAN AMEUR AND HARRISON DONNELLY, *THE ROKEBY VENUS*

HANAN

The other big action I was involved in was damaging the glass shield protecting *The Rokeby Venus* painting in the National Gallery – together with Harrison Donnelly. We chose it deliberately to follow in the footsteps of the action taken by the Suffragette Mary Richardson, who slashed the painting with a butcher's meat cleaver in 1914. Our trial for this is coming up in June 2025. (By the way, I think about this as something done by the Suffragettes collectively, rather than by Mary Richardson individually. She turned out to be a really nasty piece of work, as head of the women's section of the British Union of Fascists!)

I often thought about what we would have done if the glass shield hadn't been there, and I do think we would still have attacked the painting itself. I know that would have enraged people even more than the action did anyway, but even a painting as historically significant as that is still just an inanimate object, and if we go on risking complete climate breakdown, then people are not going to be worrying about paintings in museums.

We talk all the time about whether the line can be drawn between 'acceptable violence' and 'unacceptable violence' when it comes to property – obviously, Just Stop Oil will *never* be involved in violence against people, but it's not an easy line to draw when thinking about the extent of any criminal damage.

All I can really say is that it's deeply annoying that I have to go and smash up a screen in front of a painting and get myself arrested just to persuade more people to start listening to what the scientists are saying about the climate crisis. That's the bottom line here: why are so few people today, even now, focused on that scientific advice?

HARRISON

We obviously chose *The Rokeby Venus* painting for highly symbolic reasons. We really wanted people to be able to think about the links between what the Suffragettes were doing more than a hundred years ago to ensure that women had a vote — and particularly the action taken by Mary Richardson in slashing that painting — and what we're doing today to confront the climate crisis. It such a dramatic thing that she did at the time, forcing people to think about the injustice being addressed by the Suffragettes, and we wanted to force people to think about the climate crisis in the same way. Would we have used a butcher's meat cleaver, as Mary Richardson did, instead of our little safety hammers, if there hadn't been a screen in front of the painting today? Good question!

The Suffragettes provide some of the evidence that non-violent action can work, and that's why so many people in Just Stop Oil see them as a source of real inspiration. It's difficult for us that much of their campaigning later on condoned the use of violence against property, and much more violent tactics in general.

But for Just Stop Oil, we know that violence is not the answer for us. What we're trying to do is to show compassion for all those people who are already experiencing terrible violence because of the climate crisis — and for many, many more who will experience that violence in the future. Food shortages, droughts, wildfires — this is all about the violence being done to the Earth and to its people. And you can't fight fire with fire. You've got to fight fire with water; you've got to fight violence with compassion. That's absolutely at the heart of everything we do.

This is really hard to explain, but I felt a lot of this very intensely the night before our action. I really didn't want to have to do it, because I knew perfectly well that it could influence the whole of the rest of my life, that it might mean I wouldn't be able to become a university lecturer, for example. But at the same time, I felt that I really didn't have a choice — I absolutely *had* to do it. Otherwise, we were all going to end up in a world where being a university lecturer, or anything else, for that matter, will mean literally nothing. So I was very, very scared, given that uncertainty. But because I knew I didn't really have a choice, there was also this sense of relief.

Richardson's action in 1914 caused all public galleries to be closed. Emmeline Pankhurst was talking openly of the need for 'further guerrilla warfare', even as their goal seemed as far away as ever. But it turned out to be almost the final protest before the onset of the First World War. The Suffragettes declared 'a truce', and all suffrage prisoners were unconditionally released. Emmeline Pankhurst became a staunch supporter of the war effort; others campaigned against it for the entire four-year duration.

It's impossible to say what would have happened with the campaign if such a traumatic cataclysm hadn't changed everything. At the end of the war, a limited victory was secured for women's suffrage through the 1918 Representation of the People Act, which abolished nearly all property qualifications for men and enfranchised women over the age of thirty who met certain minimum property qualifications.

In all the conversations I've had, I never detected the slightest intimation of anyone wanting to escalate from Just Stop Oil's original position – a readiness to commit limited criminal damage, breaking windows and vandalising petrol pumps, in pursuit of its objectives – to 'blowing things up'. Andreas Malm's *How to Blow Up a Pipeline*, and the film of the same name, is widely discussed, but resorting to more extreme forms of violence against property, which inevitably entails a clear risk of people being injured or killed, was never in the Just Stop Oil playbook.

However, it seems inevitable that new organisations will emerge that turn their back on non-violence and opt instead for more violent tactics, including eco-sabotage. In January 2025, a group called Shut the System set out to disrupt internet services for insurance companies in the City of London by cutting some underground cables, promising to 'kick-start a new phase of the climate activist movement, aiming to shut down key actors in the fossil-fuel economy'. By all accounts, the disruption was minimal, and Shut the System subsequently acknowledged that 'there is obviously a learning curve to these things'.

The debate about the morality of using violent methods to advance political ends is always with us, especially around that contested grey area as regards sabotage and the destruction of property. I was struck

'My first meeting with Just Stop Oil was a real game-changer. I literally walked into the room feeling one thing about arrestable action and walked out feeling quite different. And that came from hearing the story about the Freedom Riders (see p. 167) and how they were willing to sacrifice everything, writing their wills before they undertook those bus rides – it just put everything into perspective for me. I'm white, middle class, with a lot of security in my life. So civil disobedience is something I can absolutely do. That moment got me over the fear barrier in making that decision about risking arrest – and that actually felt like a massive relief.

'It didn't mean that I wasn't afraid when it came to crossing that line – and the initial action that I took at the oil terminal was both physically demanding and absolutely terrifying. And so far removed from anything I'd ever contemplated in my life before! But there is no way I was backing out, having got that far. In comparison to the oil terminal action, climbing the gantry over the M25 was relatively simple. I can still remember, as I climbed up that ladder, a sense of how astonishingly easy it was to go from being an ordinary pedestrian to being "a public nuisance".

'Martin Luther King always knew that there would be no convenient endpoint, where everything would be sorted. The struggle for "beloved community", as he called it, would be an ongoing journey. As people working for justice today, we need to understand that we've taken up the mantle in our lifetimes, and there will always be others continuing that work after us.

'The way in which we do that work is also critically important. Nonviolence isn't just a tactic – there's something so much deeper than that, pointing always to the need to keep ethics at the heart of everything we do. I've come to understand nonviolence as an holistic framework, a way of being in the world – the way we live, how we need to do both inner work and outer work, how we must focus on justice as fundamentally intersectional. There are so many forces at work in society today that rely on division and hatred, and our work has to be underpinned by a constant determination to combat those things.'

Cressie

by these lines from Michael Albert, in the hugely influential book *This Is An Uprising*, by Mark Engler and Paul Engler:

> It's really quite simple. The state has a monopoly of violence. What that means is that there is no way for the public, particularly in developed societies, to compete on the field of violence with their own governments. That ought to be obvious. Our strong suit is information, FAQs, justice, disobedience and especially numbers. Their strong suit is lying and especially exerting military power.

Which is why, at this particular moment, the leadership of Martin Luther King speaks as powerfully to young Just Stop Oil activists as that of the redoubtable Emmeline Pankhurst.

THE CIVIL RIGHTS MOVEMENT

By 1961, the Civil Rights Movement was facing a crisis. It was six years since Rosa Parks had led the bus boycott in Montgomery, but despite a Supreme Court decision in 1960 ordering the desegregation of interstate transit systems, nothing had really shifted since then, and in most of the Southern states, seating on buses and in bus stations remained segregated by race.

So, in 1961, the Congress of Racial Equality initiated its Freedom Riders campaign, with white and African Americans sitting together on buses simply to uphold the Supreme Court's decision. These actions resulted in appalling violence in places like Birmingham, Alabama, often co-ordinated by local Ku Klux Klan groups, only after which would the police move in to arrest the Riders for trespass, unlawful assembly and other trumped-up charges. Initially, President Kennedy and Attorney General Robert Kennedy called for a 'cooling-off period' – apparently they were appalled at the prospect of the USA being embarrassed on the

'It helps to think about past examples of civil disobedience, which tell us that one needs only a small percentage of people to undertake those acts of rebellion – with "just a roll of the dice" determining the success or failure. History tells us that with a small number of people out in front making stuff happen, everything can then shift. But you never know when something is going to happen, and if you're not trying, then perhaps that moment is never going to come!'

George

'I prefer to talk about "peaceful direct action" rather than "Non-Violent Direct Action". Simply using the word "violent" puts it in people's heads, even though it's negated. And "violent" is a bad word. When violence is talked about, we've already tipped over into despair. "Peaceful" implies that there is still hope. And we cannot be in despair – as long as there is something left to save, then we must try to save it.

'This was such an important part of the action at Stonehenge. The whole point was to jolt people into action: you see something you love being "attacked" in that way – in this case, Stonehenge – and you react with outrage, demanding that something should be done to protect it. Well, isn't that the same thing, an allegory for what's happening to our planet? Isn't that how we should all be feeling about the planet, about the things we love? We need the same kind of response about the whole of life on Earth – so I hope it made some people think about that, drawing parallels.

The hostile response didn't really surprise me. But it's horrendous that most of the media won't even give Just Stop Oil activists a platform to explain why they're doing what they're doing. It just labels people like me as "eco-zealots", but they don't really want to know anything about us, about what motivates us.'

Niamh

world stage by such racist violence. The Freedom Riders declined, and the Rides kept on through the summer, involving around 500 mostly young people, with equal participation from black and white citizens. There was no let-up in the violence.

Soon after that, Martin Luther King's own organisation, the Southern Christian Leadership Conference (SCLC), came to the conclusion that there was no alternative to this kind of direct confrontation, and set out to engineer a major crisis in Birmingham in April 1963. The chosen goal was to persuade the Birmingham business community to start desegregating all their stores and restaurants – not least help create additional prosperity for the city as a whole.

People forget just how controversial this was at the time, as I touched on in Chapter 1 with reference to the 'Radical Flank effect'. Martin Luther King himself was seen as a 'dangerously polarising' figure by more mainstream organisations like the National Association for the Advancement of Colored People, supported by established black churches and any number of organisations made up of white liberals, all purporting to be 100 per cent committed to securing civil rights for African Americans, as they had been for many years. When King himself was arrested on Good Friday for kneeling down to pray in front of Birmingham City Hall, there were many urging him and the SCLC to call off their protests before serious violence broke out – 'when rights are constantly denied, a cause should be pressed in the courts and negotiations among local leaders, and not in the streets'.

Writing from Birmingham Jail, where he was in solitary confinement, King articulated a sentiment that has underpinned all NVDA campaigning since that time:

> You are exactly right in your call for negotiation. Indeed, this is the purpose of direct action. Non-violent direct action seeks to create such a crisis and establish such creative tension that a community that has constantly refused to negotiate is forced to confront the issue.

'One thing that immediately impressed me about Just Stop Oil was the emphasis on non-violence training. So many lessons have been learned from the Civil Rights Movement in the USA, where they went to incredible lengths to prepare campaigners for the kind of violence that might be used against them – as it so often was – to remain completely calm, non-violent and non-reactive. This was one of the ways in which they built up an extraordinarily strong community, underpinned by their Christian faith. JSO doesn't draw on that kind of spirituality, but the solidarity of the movement, and of so many other campaigns where non-violence has worked, has been enormously influential.

'It's weird to hear oneself described as a "terrorist" or "eco-zealot"; people seem to forget the role that non-violence has played in other progressive movements. Drawing parallels with the Suffragettes, who are now held in such high esteem by so many people, celebrated as champions of democracy, people conveniently forget that the Suffragettes actually slashed paintings rather than just throwing some soup at them! It's crazy!

'And then there's the line of attack that says that all this is just a "slippery slope", leading to terrorism – as in Lord Walney's suggestion (see page 195) that "they may not be using violence now, but there is a real risk they might use violence in the future". So you get done for the prospect that somebody might use violence in the future, even though everything that Just Stop Oil says and does is one hundred per cent based on non-violence.'

Rosa

7: TAKING DIRECT ACTION

Three weeks after his arrest, the SCLC authorised what has become known as the Children's Crusade, allowing high-school students to join the protests. The police promptly increased the level of violence, using batons, high-pressure water hoses and police dogs against these young protesters. This outraged Birmingham's more conservative black community, who then came out on the streets in support. On 7 May, more than 1,000 people were arrested. On 10 May, business leaders agreed to start talks with the SCLC on a timetable for desegregation. The violence ended, and those who had been imprisoned were soon released.

The protests continued elsewhere. During that summer, there were more than 1,000 demonstrations across the South, with upwards of 20,000 people arrested. In 1964, President Johnson signed the Civil Rights Act. Most commentators agreed this would not have happened without the Freedom Riders or the events in Birmingham. The idea of 'confrontation as strategy' had been seen to be vindicated.

NVDA: SUCCESSES AND FAILURES

There are countless examples of when NVDA has been of critical significance in achieving progressive social change. Many of these examples feature prominently in 'the constellation', as one of my interviewees described it, of influential predecessors which inspired Just Stop Oil activists: Mahatma Gandhi's 'salt campaign' in India in 1931; the sustained campaigns against apartheid in South Africa and the UK; the ousting of President Milošević in Yugoslavia in 2000, driven by the vibrant, highly creative non-violent campaigning of a civic youth movement called Otpor!; the hugely successful protests in Poland, where the Solidarity movement became a household name across Europe; East Germany at the time of the collapse of the Iron Curtain at the end of the 1980s; the even more dramatic ousting of Egypt's President Mubarak after the demonstrations in Tahrir Square in 2011; the brilliant campaigning of ACT UP (AIDS Coalition to

'With my first action, it took only a few minutes of marching in the road before I felt this incredible energy – I was really properly trying to do something meaningful for the first time and felt with absolute certainty that that was where I should be. This was confirmed by the first time a white van drove past us in the road, with a man screaming "wankers" at us!

'This was the longest civil resistance campaign in the UK at that time, lasting thirteen weeks. It was certainly a turning point for me, as I'd never thought I'd be able to do something like that. I'd always thought I wouldn't be brave enough or strong enough. Because I'm autistic, I sometimes don't leave the house for days on end.

'Six months later, in November, I first got arrested. I had got to know a lot of people in Just Stop Oil, and taken inspiration from them, finding joy in the company of like-minded people. Up until then, the conversation with myself had been all about what difference could I make, on my own, what direct impact could I have? By November, it was more along the lines of what could *we*, as a group of people, do to make a difference.'

Jacob

'When we were sentenced in court [27/5/2025], we each raised signs saying, "Billions will die". The science is clear ... I consider the facts to be so alarming, so stark, so utterly heartbreaking that disruption to everyday life is warranted. And I've spent each day in custody questioning why others equipped with the same knowledge as I have do not feel the duty to act in the same way that I do.

'The judge agreed that we acted on our conscience, but for sentencing, he wanted to see remorse. But how can it be possible to take part in an act of conscience and then show remorse? How could I be morally compelled to take action one week, and then be filled with regret for acting the next?'

Indigo

Unleash Power) in the USA in the 1980s, forcing politicians to take urgent action on the AIDS crisis.

There are also, unfortunately, many examples of when NVDA has ultimately failed to achieve even minimal social change, including Occupy Wall Street, Occupy London, in 2011, and many other Occupy campaigns protesting against austerity measures introduced after the financial crash in 2008; also the ill-fated 2014 Umbrella Movement in Hong Kong, demanding more transparent elections, but which culminated in the final crushing of democracy in Hong Kong ten years later, in 2024, with the sentencing of forty-five pro-democracy protesters.

These were all 'hot topics' amongst Just Stop Oil campaigners. The age-old debate about the differences between, on the one hand, established organisations set up and sustained, often over decades, to achieve gradual social change, and, on the other, disruptive, momentum-driven mass mobilisation movements seeking to break through layers of inertia and the endless barriers created by vested interests, is as live today as it's ever been. And when it comes to the climate crisis, as I explained in Chapter 1, the competing theories of change that lie behind these two models are both being exposed to scrutiny as never before – quite simply because neither has, as yet, delivered anything even vaguely resembling the kind of change that the crisis demands.

I know of very few people in the mainstream climate movement who feel we are still on the right track. Painful though it is, they can see the gulf between what the science of climate change is telling us, right now, and the collective political response. There is no narrowing of that gulf. Indeed, with the massive backlash now underway, as highlighted in Chapter 4, it is clearly only growing wider.

That does not mean, however, that these mainstream organisations are about to change their ways. There are many self-styled 'stubborn optimists' or 'incurable solutionists' who are still persuaded that their mainstream theory of change remains fundamentally sound, that all we need to do now is to double down on getting politicians on board, and that this will, eventually, achieve what is needed – as it has on so many previous occasions with other campaigns. It's also true that they have no idea what else to do, and are either apprehensive about or actively hostile to the alternative theory of change embraced so uncompromisingly by Just Stop Oil throughout three years of campaigning.

Over the last year or so, I've been taken aback by the number of mainstream colleagues who have little respect for Just Stop Oil, and very little sympathy for those activists who have been arrested and, in some cases, imprisoned. As I've mentioned before, I'm endlessly told that not only are such tactics ineffective, achieving nothing, but that they're clearly counterproductive, alienating more people than they're attracting into active campaigning work.

Personally, I don't believe that to be the case – not least because this idea of alienation of potential sympathisers is one relentlessly pursued by the UK's right-wing media, manipulating whatever levels of anxiety or anger that might genuinely exist, and almost obsessively demonising Just Stop Oil activists as eco-zealots, hypocrites or attention-seeking extremists. Without the media's poisonous campaigning, I doubt we would have seen so many Conservative politicians seek to bring climate change into their woke-busting culture wars, with Liz Truss, Rishi Sunak and now Kemi Badenoch all enthusiastically distancing themselves from the cross-party consensus that so crucially brought the Climate Change Act into existence back in 2008, let alone the introduction of ever-more draconian legislation, which I will address in the next chapter.

It is true, however, that Just Stop Oil's confrontational use of NVDA did not shift the needle in the way that it was once hoped it might. Targets for the reduction of greenhouse gas emissions have been tightened; the rhetoric has often been ramped up; and a few policies have been adopted that can reasonably be interpreted as an indication of more progressive things to come. For all that, it became a widely held view amongst Just Stop Oil campaigners that they were being punished, through the new anti-protest laws and draconian sentences, precisely because they were so effective in highlighting the consequences of the Climate Emergency. But the needle still hasn't shifted.

Perhaps we shouldn't be too surprised by this. In August 2024, the Mercator Research Institute on Global Commons and Climate Change published a paper assessing the impact of 1,500 climate policies between 1998 and 2022 in forty-one different countries. Only sixty-three of those policy interventions could be demonstrated to have achieved any 'substantive reduction in emissions'. This tells me that politicians the world over still aren't really serious about addressing the Climate Emergency.

So when it comes to the two competing and contrasting theories of social change, I still feel the 'direct action disrupters' have a much better case to make than the 'double-downers' – and that's before we start to think about some of the moral implications entailed in both approaches, knowing how little time we have left before it becomes too late to avoid irreversible climate breakdown.

This is why these lines from the inspirational abolitionist Frederick Douglass in 1897 are much loved by NVDA campaigners:

> If there is no struggle, there is no progress. Those who profess to favour freedom, and yet deprecate agitation, are men who want crops without ploughing up the ground. They want rain without thunder and lightning. They want the ocean without the awful roar of its many waters.

JACOB PINES

I'm a twenty-six-year-old energy specialist working for a renewable energy supplier. I'm keen to have a greater impact on the multi-crisis we're currently facing.

INVOLVEMENT
I took part in Just Stop Oil's slow march campaign from the end of April to November 2023. I was arrested twice in November London marches. I took part in the first nationwide campaign with Defend Our Juries in September 2023.

I was raided at my home address by the police in June 2024 as part of a police operation in anticipation of protests at airports later in the year.

MOTIVATION
There's no one thing that drives me towards taking action. The reasons are countless, and every day I discover more: a family member or friend is pregnant or happily engaged; someone tells me their dreams and aspirations; a new species is discovered or found to not be extinct, as was once previously believed. I'm always learning about the wider world and how it is even more interconnected than I already knew, and how every decision we make affects others. Everything that exists and will be depends upon a stable climate. Without that, nothing but suffering can materialise. I'm going to die one day, after which nothing that I owned or anything that was said about me will matter. The only thing that will matter will be how much I did to ensure life can continue.

INSPIRATION
A woman I fell for whilst we were engaged in civil resistance together last year: she's been through so much and yet she puts herself in harm's way and carries

what feels like the weight of the world on her shoulders for what she loves and believes in. She is strong, smart, fierce, powerful and always leads with compassion. I won't name her, but if she ever reads this, she knows who she is.

IN NATURE
I don't personally feel too much of a connection to any one location in nature. I've been much more personally involved with man-made society, which, whilst artificial, is still a product of humans, who are themselves a part of nature.

I would say my connection to nature stems from my desire to lead with compassion in everything I do. It's natural to be kind; and I believe to be empathetic is to be curious, and perhaps later in life, wise. To be driven by fear, envy, greed and hatred is to blindly rip ourselves from nature and reap the destruction we bring to bear on our surroundings and ourselves. Love for oneself, friends and family and any other life that we come across, can recentre us and connect us with the natural world.

QUOTATION
'Never doubt that a small group of thoughtful, committed citizens can change the world; indeed, it's the only thing that ever has.' Margaret Mead

RESOURCES
The Climate Book, by Greta Thunberg, is a brilliant and approachable resource, from experts in many different fields on why things are happening, what has to be done and how all of us – you included – can help to bring about that change. It also includes excerpts written by Thunberg herself, including the message that not all of the ideas that are necessary to bring about the change we need have been thought up yet, and that you can and should push to be part of the next group of people who help to bring the next piece of the puzzle into place.

WHAT LIES AHEAD?
I would hope that we as a nation, and as an international community, do everything we can to bring about climate justice, with all of the social aspects that come with it. I want to ensure that no one is left behind as we ascend from

simply being a species that exploits the world we live in to one that acts as its custodian. I don't believe that I'll live to see an egalitarian utopia that could rise from the ruins of capitalism, but I hope to act as a stepping stone towards it. Failing that, given how dire things already are and are likely to become, I will endeavour to ensure that there is a semblance of global order so that we can meet head on the challenges that will inevitably arise.

NIAMH LYNCH

I'm twenty-two and in my third year studying geography at Oxford, specialising in ecology, conservation and climatology. I spent last summer on islands doing fieldwork for my dissertation, looking at how sea-level rise is affecting coastal wading birds like the ringed plover and oystercatcher. I was living in my teeny tent for a month and a half, swimming in the sea every morning and cooking on my little Trangia – it was the best! I hope one day I'll be able to study the birds I love so much without fear that they'll disappear forever, that that may be the last moment I get to spend time with them.

INVOLVEMENT

I've been involved with XR since I was at school, and with Just Stop Oil since I started at Oxford. In my first year, I slow marched in London and in my second year, together with Rajan Naidu, I carried out the action at Stonehenge, just before the solstice on 19 June 2024. I was arrested, then released on bail. Our case has been referred to the Crown Court – and the trial is expected to be held in the summer.

MOTIVATION

Activism is about recognising that a problem exists and then trying to do something about it. It's about acknowledging that something is wrong, believing it can be better and that it doesn't have to be this way, and then doing something to fix it, however small – and not just by getting arrested!

In that way, I've always been an activist. My parents raised me with this mentality. My upbringing emphasised gratitude, to not waste resources, appreciating that these things come from the Earth. Food, a gift from the rain, the sun and the soil, should be treasured.

My parents and my grandfather instilled in me a deep, deep love for the natural world. If you love something and see it being hurt, you want to do something about it. It's so natural. It is this deep love of the Earth and all of the living beings we share it with that motivates me. I'm fuelled by love, not anger; I have an absolute desire to protect and preserve the things that mean so much to me.

INSPIRATION

It's about a community of people, from my family and friends to everyone I've met in the world of climate activism – everyone who cares and gives some of their time to fighting for change and believing the world can be better – they're the ones who inspire me.

I spent a few months when I left school working on a Royal Society for the Protection of Birds (RSPB) reserve, monitoring hen harriers and looking after a herd of Belted Galloway cows. One of the wardens I worked with was a hunt saboteur in his youth. He told me incredible stories of immense bravery and courage, standing up for and doing everything within your power to protect what you love.

Since our Stonehenge action last summer, Rajan has been a constant source of inspiration to me. His unwavering willingness to act for the greater good, putting the needs of others above his own, despite his own fears and worries, is truly incredible. A special person indeed.

IN NATURE
For me being outside just feels right, being inside feels so very wrong. It always has. If I don't feel the sun on my cheeks or the wind in my hair I just don't feel real, I feel like I haven't lived that day.

I'm most alive and at peace on our family farm, watching the swallows careering over the hay meadows, or romping around the fields behind my village, watching kestrels and barn owls patrolling and hares boxing, or finding lapwing nests in the stubble.

Whilst I'm in Oxford, Boundary Brook is my little haven. It's a wee nature reserve in east Oxford, a tiny island of peace, cut off from the city. In amongst its ash groves and hedges of guelder rose, I've learnt to coppice hazel, weave willow and find brown hairstreak butterfly eggs.

QUOTATION
When Trump was re-elected in the US everyone who cares about protecting our world was shaken. Cutting through the atmosphere of despair and panic, Chris Packham wrote: 'I'm not going to give up on the beautiful and the good. The grip on my dreams just got tighter.' It's saved on my laptop screen and written on the wall in my room. It reminds me that despite the chaos, we know what's wrong and where we're going.

WHAT LIES AHEAD?
I'm an inherently positive person and just want things to be better, for every living being on our planet. Every step in the right direction, however small, is so important. We must be ambitious, determined and unwavering in our push for transformative change, celebrating the small things, but always asking for more. We need a world without fossil fuels, where we live in harmony with nature, and living beings are no longer needlessly dying. We should be satisfied with nothing less.

8: THE WEIGHT OF THE LAW

'Is it justice when the powerful are not held to account, and ordinary citizens are prosecuted when they then call them out?'
Cressie Gethin

FOR ANY young person contemplating getting involved in NVDA today, it's a very different picture from the one when XR first took to the streets of London back in November 2018. Six years on, people's basic right to protest has been dramatically constrained, policing has become far more oppressive, levels of surveillance have been hugely ramped up, court proceedings made more and more restrictive, and sentencing has become extraordinarily harsh.

Things began to shift right from the start of this period. Five bridges were blocked by XR in November 2018, and there were many arrests. But the mood was almost celebratory; the police were definitely in full-on friendly mode – at least to begin with. But The Home Office and the Metropolitan Police had plenty of time to think about tactics before the next London rebellion in April 2019, when five major traffic intersections in London were blocked. Roger Hallam, one of the co-founders of XR, had declared that one of their aims was 'to overwhelm the court system', with at least 400 needing to be in prison and up to 3,000 arrested'. The mood was very different. More than 1,130 arrests were made. And in October 2019, third time around, 1,800 were arrested. However, most were conditionally discharged or given suspended or community sentences.

In fact, many XR defendants were found not guilty even when they did not deny having done what they had been charged with. In April 2021, for instance, a group of XR activists who had caused some damage to Shell's London HQ were found not guilty by the jury – even though

COLE MACDONALD

One thing I do feel very strongly is that any action has to be directly linked to the cause of the problem – our continuing dependence on fossil fuels and the fact that some people are more responsible for that than others. Politically and environmentally, I've always hated the use of private jets and what the growing number of people owning private jets tells us about society at large. So it was a good fit for me to undertake my first JSO action spraying private jets with orange paint.

I had very mixed emotions during the moments before our action started. On the one hand, I really couldn't imagine how I was going to do it and was scared shitless. On the other, I really couldn't imagine how I could ever *not* do it! I was holding both these emotions at the same time.

After I was arrested, I ended up spending three days in solitary confinement in a police cell before appearing in court. It was definitely one of the weirdest and most upsetting experiences of my life, but also one of the most transformative. It was really scary, in all sorts of ways, and because there was no daylight, I never knew what the time was. But I do remember it was freezing cold, that the vegan lasagne was absolutely disgusting, and that I spent a lot of time walking around my cell singing, 'I wish there was no prisons'! I got a real buzz when I found a Just Stop Oil skull, as in our logo, drawn in the grouting between the wall tiles – 'I'm not alone,' was all I could think.

I was then remanded to prison for two weeks – I'd known this was a possibility, but as a twenty-two-year old, with no prior convictions, I guess I hadn't expected to be remanded in that way. And then, unexpectedly, I was released on bail – just as I was getting used to the idea that I was going to be on remand for a long time.

But because of the bail conditions, I feel I'm still imprisoned mentally. I have to sign in every day, which is virtually unheard of; I have to wear an ankle tag, which I find physically violating, as if they really do 'own you'; and I'm not allowed to have any contact with Jen, who I did the action with. And I'm stuck up here in Sheffield, even though I'm paying rent in Brighton!

My trial isn't until September 2025, and things keep getting pushed back all the time. But the sentences imposed on the Whole Truth Five have made me very fearful about the sentences that we are likely to get.

8: THE WEIGHT OF THE LAW

the judge had explicitly directed them that the defendants had no defence in law just because they were so concerned about the climate crisis. In January 2022, three XR campaigners were found not guilty of causing significant disruption at Shadwell Station – even though they clearly had. Two of them were Church of England priests, including the inspirational Sue Parfitt who was arrested again in May 2024 for damage done to the protective glass around the Magna Carta in The British Library.

And in November 2023, nine members of XR who had admitted that they caused £500,000 of criminal damage to HSBC's HQ – HSBC being one of the worst offenders when it comes to continued funding of fossil fuels – were also found not guilty. It took the jury just two hours to come to that conclusion, mindful no doubt of the time-honoured principle in English law that 'jurors have the absolute right to decide their verdict according to their conscience'. More on that in a moment!

However, by the start of 2020, XR was in some disarray. There had been massive controversy about another train protest, at Canning Town station in October 2019, where two XR activists had retaliated when they found themselves being assaulted by irate commuters. And then came Covid, which had a devastating effect on XR. Eighteen months on, in August 2021, it tried to get its mojo back with its 'Impossible Rebellion', but this was widely judged to have been a very damp squib indeed. In December 2022, it issued its high-profile 'We Quit' declaration: 'We have made a controversial resolution to temporarily shift away from public disruption as a primary tactic, whilst we still recognise the power of disruption to raise the alarm.'

In April 2023, it organised 'The Big One' outside the House of Commons, including speeches, stalls, artists and a polite die-in outside the Treasury, which I enjoyed being part of! All good stuff, with around 100,000 attending over two days, but almost completely ignored by the media. XR has now reinvented itself in the UK, spearheading some innovative campaigning against the insurance industry. It's now something of a hybrid when it comes to its revised theory of change: get seriously good at conventional campaigning tactics, as used by mainstream

'I was shocked by those sentences handed out to the Whole Truth Five, some of whom are good friends. This felt really different, as if we were suddenly living in Russia rather than in the UK. How can it be that people on a Zoom call can be arrested for a discussion about a protest in the future, which might not even have gone ahead? It shows the lengths to which government will now go to crush peaceful protest.'

Harrison

'I have always been extremely careful about not getting involved in actions that would seriously jeopardise my future career. I'm not the kind of person who rushes into things anyway. But I know there's nothing unique about me and the contribution I might be able to make as a scientist – there have been bloody loads of climate scientists over the last fifty years! One of the things that continues to make me angry is the way in which so much of that scientific endeavour has simply been ignored, with people now paying an increasingly heavy price for the decisions the politicians have taken *despite* the science.'

Ollie

'Community matters. I have a friend in Palestine Action who was really badly treated by the police, yet ended up feeling guilty about letting this get to them. For us, there's always going to be a group of people who will look after you, who will go through things emotionally with you, even help practically with food and housing. They're going to be there for you, even if the actions you take end up brutalising you emotionally.

'Immunity from that brutality comes from the community we share.'

Daniel K.

8: THE WEIGHT OF THE LAW

organisations, whilst retaining a vibrant, noisy presence on the streets – but not going for major disruption or mass arrests.

The use of NVDA in climate campaigning was taken up, in September 2021, by Insulate Britain, which had a big impact for a very short period of time. It was gone by February 2022. The disconnect between its campaign goals – to pressure the UK Government into addressing chronic fuel poverty and the dreadful housing conditions that still exist in the UK – and its adopted tactics (protesters gluing themselves to motorway slip roads) – was difficult for the public to connect with.

Insulate Britain was specifically planned as a short-term campaign, with a view to ensuring that enough activists were arrested and imprisoned to embarrass Prime Minister Boris Johnson when he hosted COP26 in Glasgow at the end of that year. However, the police at that time decided not to remand activists in prison, even after multiple arrests, and started using injunctions as a deterrent rather than the criminal law. The crisis in the criminal justice system means that many Insulate Britain activists will not be tried until 2026.

Insulate Britain was immediately succeeded by Just Stop Oil in February 2022 (see interview extract from Indigo Rumbelow on page 27), causing the Home Secretary and other Government Ministers to give voice to increasingly vituperative and often completely over-the-top comments about 'unacceptable criminality'. But it still came as a shock, in April 2023, when Morgan Trowland and Marcus Decker got the longest sentences in modern history for causing a public nuisance – three years and two years and seven months respectively – for displaying a Just Stop Oil banner on the Queen Elizabeth II Bridge over the Dartford Crossing, causing significant disruption to traffic.

Come Christmas 2024, nineteen Just Stop Oil activists found themselves in prison, nine having received lengthy sentences, with a further ten on remand. It was apparent by then that keeping defendants in prison on remand, rather than releasing them on bail, was very much part of the overall crackdown. Even those who were released on bail found themselves having to comply with extraordinarily restrictive conditions.

'The M25 gantry protest was much more serious than anything I'd done before – and somehow became a symbol of the apparent irresponsibility of us "climate extremists". We ended up having to deal both with a criminal charge and a prosecution in the civil courts after an injunction from the Highways Agency. Quite a lot of people are having to deal with this kind of double jeopardy.

'We've had to deal with all the new legislation coming out of the Conservatives, stacking the odds against protesters, deterring people who want to get involved, and stacking the odds in the judicial system itself – with judges telling defendants that they were not allowed to talk about the climate crisis and why they had felt the need to do something about it.'

George

'It's almost impossible to explain to people quite how disorganised and chaotic the criminal justice system in the UK is today. They don't have enough judges; they don't have enough court space; and the infrastructure needed to support so many trials is absolutely shambolic. Time after time, people turn up hours late for their own trial, simply because the transport wasn't available. So you have judges tapping their feet all day – a huge waste of public money!

'And it's now very clear how the 2023 Public Order Act has changed things. If there were judges who were previously sitting on the fence about balancing defendants' rights to protest and freedom of speech with the need to maintain public order, they're now clearly under a mandate to get off that fence! Deterrent sentences have become the norm; those rights have been dramatically restricted.

'I was very involved in the "Kill the Bill" campaign, and there were millions of people who were very concerned at its implications for freedom here in the UK. But I don't think many people realised then quite how bad its impact would be.'

Indigo

8: THE WEIGHT OF THE LAW

ACTS OF OPPRESSION

When people talk about 'the UK's draconian new laws regarding limitations on the right to protest' they're referring to two major new pieces of legislation:

1. The Police, Crime, Sentencing and Courts Act (2022)
This sprawling Act includes a number of measures to restrict protest, sweeping new powers for the police, and, almost as an aside, new sanctions against Gypsy, Roma and Traveller communities. Despite significant opposition, including a petition signed by almost a million people, the Tory Government bulldozed it through Parliament, rejecting a number of attempts by the House of Lords to remove some of its more egregiously repressive measures.

2. The Public Order Act (2023)
When the new Police Act still proved to be insufficiently repressive, as Just Stop Oil continued with various actions and campaigns, the Tory Government came back with another set of measures, updating the 1986 Public Order Act, including, in Section 7, a definition of 'national infrastructure' specifically designed to deter Just Stop Oil from some of its actions. Protesters who choose to 'lock on' to an object or to another person, with some form of adhesive or handcuffs, will now face up to four-and-a-half years in prison. This is more than twice as long as the maximum sentence for racially aggravated assault in the UK.

It's worth reminding ourselves just how 'historic', in a bad way, these two Acts are – in the words of George Monbiot back in July 2022:

> Protest is not, as governments like ours seek to portray it, a political luxury. It is the bedrock of democracy. Without it, scarcely any of the democratic rights we now enjoy would exist: the universal franchise; votes for women; civil rights; equality before the law; legal same-sex relationships; progressive taxation; fair

conditions of employment; public services and a social safety net. Why do governments want to ban protest? Because it's effective.

Then Home Secretary Suella Braverman did not get everything she wanted in 2023. Having failed to persuade the House of Lords to leave it up to Ministers to define what constituted 'serious disruption' caused by any protest, she then used regulatory powers to amend the earlier 1986 Public Order Act, bypassing Parliament to achieve the same goal. In May 2024, judges in the High Court, ruling on an action brought by civil-rights campaigners at Liberty, found that she had acted unlawfully. The Tory Government decided to appeal that ruling. Almost unbelievably, the new Labour Home Secretary, Yvette Cooper, decided to continue with the Appeal rather than withdraw it. Libertarian instincts run very shallow indeed in today's Labour Party.

At the same time as the entire legal framework around the right to protest was being transformed, the judiciary also began to fall in line with government policy. It is hardly coincidental that it was at this very point that certain judges set out to curtail the rights of defendants in courts, barring them from making use of certain 'defences in law', including being able to explain their deep concerns about the climate crisis as a defence for the actions they had taken, sometimes seeking to prevent any mention of climate change at all, and refusing requests from defendants to allow expert witnesses to speak on their behalf. In 2023, Judge Silas Reid imprisoned three people for defying his ban on mentioning the words 'climate change' or 'fuel poverty' in his courtroom.

DEFEND OUR JURIES

Suppressing the rights of defendants is one thing; suppressing the rights of juries is another altogether. Juries are at the heart of our criminal justice system, and individual jurors are free to exercise their moral conscience in determining a verdict against defendants. It had become an

8: THE WEIGHT OF THE LAW

extreme embarrassment to Ministers that juries were repeatedly reaching not guilty verdicts in prosecutions involving XR and Just Stop Oil activists, undermining their propagandistic rhetoric that 'the general public supports the crackdown on protest, and on extremist campaigns that go out of their way to disrupt ordinary people's lives'.

As was the case, in June 2024, when three Just Stop Oil activists – Nathan McGovern, Rosa Sharkey and Louis Hawkins – were found not guilty of criminal damage in their trial at Guildford Crown Court, even though they admitted having smashed the display glass on petrol pumps and spraying them with orange paint. Twelve ordinary members of the public completely ignored the presiding judge's instruction to disregard the defendants' motivation.

In March 2023, outside the Inner London Crown Court where Insulate Britain protesters were being prosecuted for a roadblock on the M25, Trudi Warner, a retired social worker, held up a sign which read simply:

> Jurors:
> You have an
> absolute right
> to acquit a defendant
> according to your
> conscience.

Arcane though this may be – this is the UK, after all! – that wording goes back to a case in 1670, known as Bushel's Case, in which a jury refused to find the defendants in front of them guilty despite having been repeatedly instructed to do so by the judge. The very same words can be found inscribed on a marble plaque in the entrance hall of the Old Bailey.

Silas Reid, the judge presiding over this particular case, had Trudi Warner arrested, referring her to the Attorney General to consider charging her for contempt of court, with a potential sentence of up to two years. The Attorney General duly obliged, and in April 2024, after many legal twists and turns, the Solicitor General found himself

'This is not just about fossil fuels. The Government hasn't taken away only Just Stop Oil's right to protest – they've taken away everyone's right to protest. We are not fighting just against oil any more, we're fighting against an oppressive regime, against the removal of our right to protest. That's why it is so important that Just Stop Oil continues to reach out to other organisations and work in solidarity.

'What links all these things? It may be a weak word, but there is such *unfairness* in the world today, throughout society. Things don't work because they're not fair; more and more people now understand how important this is. The idea of a wealth tax, for example, has become much more acceptable to people than it was before.'

Anna

'I've obviously had a lot of experience with our criminal justice system, and a lot of what is going on is frankly unbelievable, including all the delays because of backlogs in the courts.

'The criminal justice system isn't really about deterrence – if it is, it's failed entirely, given that we have the highest re-offending rates in Europe. No, it's primarily punitive, in a very political way. The new laws, and the excessive sentences being handed down – they're all part of today's culture wars about the Climate Emergency, with the criminal justice system just one more tool being deployed in these battles. But the changes in the laws about protest are highly significant: public nuisance used to be a common law offence; now it's a statutory matter, with a maximum ten-year sentence! This is crazy stuff.

'The backdrop to all of this has really changed, with the battleground now in the courts rather than on the roads. That's why an organisation like Defend Our Juries has gained such huge support recently. Deep down, people want to know that the system is working fairly, and that our basic rights are being protected.'

Paul

8: THE WEIGHT OF THE LAW

arguing in the High Court, in front of Justice Saini, that a prosecution should proceed.

Silas Reid's humiliation was matched only by the vindication of Trudi Warner when Justice Saini ruled that there was no legal basis to proceed and that government lawyers had 'mischaracterised' the evidence in claiming that Warner had acted in an 'intimidatory and abusive manner'. As someone who has held the same sign outside courts on a number of occasions since then, this sounded utterly ridiculous, knowing how clear our instructions are in how we should comport ourselves – no eye contact, no conversation. Justice Saini described Warner as a 'human billboard' given 'how little Ms Warner tries to engage with people to get their attention, or to persuade them of anything'. He dismissed the Government's case out of hand.

Unbelievably, until you remember that this was a Conservative Government, the Attorney General then decided to appeal the judgement! Before this 'performative legal cruelty' could proceed, the Government got booted out in July 2024, and the new Labour Attorney General decided in August to drop that Appeal – bringing to an end a quite extraordinary eighteen-month legal saga. Trudi Warner's statement is worth repeating here:

> It's wonderful that the right of juries to acquit according to their conscience is now unequivocally established as a legal principle in the UK. My case, and the response of people in Defend Our Juries, has shown how effective collective action can be. This is a time for courage, which we must draw from one another. We are many. They are few.

Defend Our Juries, was set up to address this deepening crisis in our judicial system, working tirelessly to support Trudi Warner and others involved in campaigns to end the jailing of people for taking peaceful action to protect life, and to ensure that anyone who has taken reasonable and proportionate measures to protect life has the opportunity to properly present that as a defence against criminal charges.

'I'm not going to pretend this is easy: preparing for the possibility of prison, whilst also holding on to the heaviness of climate collapse, alongside the family side of it, the personal side of it, the financial side of it, even the practical side of it. And then, guess what, I still have to do the washing, get something to eat and pay the rent!

'I don't think I'm necessarily more resilient than anybody else, but I have come to realise that you can build resilience, especially when you are surrounded by other people in the same position. Sharing the load makes it more doable. I know I couldn't do this at all if it weren't for the support around me.'

Ella

'It's inevitable that people now talk about us as "political prisoners" that's what we are, given the way the last Government changed the laws about the right to protest, and the way the criminal justice system has been mobilised to try to silence us.

'We need to go a whole lot further than our own cause, shining a light on the criminal justice system as a whole. What I will not be is self-victimising, because the injustice done to us is just the tip of an iceberg. Being here in prison has made me realise that we are really the luckiest of them all, and that the vast majority of women in this prison are here as a result of neglectful and oppressive political decisions – decisions that have driven them into poverty, driven them to crime, left them exposed to abuse of one kind or another, which is never taken seriously by the state, and harassed all the time by the police and the criminal justice system itself.'

Cressie

8: THE WEIGHT OF THE LAW

ARBITRARY JUDGEMENTS

Since August 2024, this crucial legal campaign has been significantly extended to draw attention to the fact that the UK now has large numbers of political prisoners in our jails, and to address the problem that a number of judges have been prominent in their abuse of the legal system both in terms of curtailing defendants' rights in court, and then in handing down extremely severe sentences.

In February 2024, the same Silas Reid, presiding over a trial of five women who broke some windows at J.P. Morgan – still the biggest investor in fossil fuels today – back in September 2021, refused to allow the defendants to explain their reasons for taking their action, on these remarkable grounds:

> The circumstances of the damage do not include any climate crisis which may or may not exist in the world at the moment ... whether climate change is as dangerous, as each of the defendants may clearly and honestly believe, or is not, is irrelevant, and does not form any part of the circumstances of the damage.

He then went even further: 'It is a criminal offence for a juror to do anything from which it can be concluded that a decision will be made on anything other than the evidence.' It is absolutely *not* a criminal offence. As I tweeted at the time: 'The priority here is simple: we must do everything we can to defend our juries. If we can't defend our juries, we can't defend our judicial system. If we can't defend our judicial system, we cannot defend the integrity of the political and governance systems on which we depend. And if we cannot defend those systems, then our democracy itself is at risk.'

However, in the Judges' Infamy Stakes, no one is the equal of Judge Christopher Hehir. In July 2024, he sentenced four Just Stop Oil protesters to four years and one – Roger Hallam, a co-founder of Just Stop Oil – to five years. Judge Hehir described the five Just Stop Oil activists as 'fanatics'. That same month, he ordered the arrest of eleven people for holding up signs outside Southwark Crown Court, one of which read simply, 'Juries deserve to hear the whole truth', despite the fact that the High Court had dismissed similar charges against Trudi Warner just a few weeks earlier.

At the end of September, it was Judge Hehir who sentenced the two Just Stop Oil activists who'd thrown tomato soup at the screen protecting Van Gogh's *Sunflowers* in the National Gallery in October 2022, Anna Holland and Phoebe Plummer, to twenty months and two years respectively. He wilfully mischaracterised their actions as a 'violent crime', equating throwing soup at a screen protecting a painting with physically assaulting a person. An appeal in January 2025 to have the sentences reduced or set aside was denied.

However oppressive the behaviour of Judges Reid and Hehir may be, this is not really about any particular individuals. It's about the judiciary as a whole. Climate scepticism of varying degrees is well known to be prevalent amongst senior judges. Baroness Carr, the most senior judge in the country, has also made statements implying that climate change is a 'matter of opinion' and not one of empirical scientific evidence. In March 2024, she delivered a ruling effectively indicating that juries should not be allowed to hear the evidence about climate change in criminal damage trials, and that defendants should not be allowed to explain their motives on the basis of that evidence. For the most senior judge in the country to describe the climate crisis as 'a matter of opinion' is truly astonishing.

As Tim Crosland of Defend Our Juries put it at the time:

> This ruling perpetuates an antagonism within the British justice system that has become impossible to ignore. On the one hand, you have the juries, who represent our communities. They keep acquitting environmental defenders when they hear the full story. And then you have some judges, paid by the state, who are taking increasingly bizarre measures to prevent juries handing down

not-guilty verdicts. But such oppressive rulings are backfiring. The public knows that the climate crisis is real and urgent. When courts suggest otherwise, the legal system loses public support, undermining the social contract and the rule of law.

HOSTILE ENVIRONMENTS

When one talks with young Just Stop Oil campaigners, they are somehow less shocked by such developments than ordinary members of the public when they first hear about them. It's been the backdrop to the whole of their campaigning lives: 'Control us, silence us, imprison us.'

What's happening inside the legal system is not an anomaly; it's one aspect of the way in which the 'Overton window' here in the UK has shifted so markedly to the right. When Suella Braverman was preparing the groundwork for what became the Police, Crime, Sentencing and Courts Act, one of her go-to inputs was a report called 'Extremism Rebellion' from the right-wing think-tank Policy Exchange. When the report first emerged, back in 2019, it seemed so off the wall as to be little more than a propagandistic diatribe, offering plenty of populist, anti-woke red meat, but little of any intellectual value – especially when one took into account that one of the principal funders of the report had been ExxonMobil. It became clear how complacent that judgement was, with many of the report's off-the-wall recommendations carried over directly into the new Act.

There are very close connections between that report (the work of the cross-bench peer Lord Walney, the Government's 'independent adviser on political violence and disruption' until January 2025, when he was sacked by Home Secretary Yvette Cooper), and some predictably toxic machinations involving Michael Gove when he was at the Home Office. He suggested then that 'the ability to use the right to protest as a reasonable lawful excuse to commit some crimes' should be removed, with a new definition of 'extremism' which would allow Ministers to ban any

'I'm not sure how long Just Stop Oil will last – essentially, we've reached the maximum number of people that we're able to recruit at the moment, with a readiness to get arrested, and we are now hitting up against all kinds of different barriers. These are incredibly dedicated people, but it's harder to recruit now, so I'm not sure what the future looks like. Perhaps Youth Demand will fill that space? That's certainly where I'm going to be doing most of my work.

'But I don't think that really matters. I'm not attached to Just Stop Oil as an organisation, but to the campaigns and the people. It's served its purpose, and has definitely had a massive impact – you can actually see how coverage about climate change spikes after different actions, and so many more people have woken up to what is really happening as a result.'

Ollie

'It's shocking the way in which the state is now exercising its power, obviously intent on crushing Just Stop Oil. The first knee-jerk reaction is "how dare they?!" Ironically, when one steps back and thinks about it a bit more, this kind of repression can be seen as a good-ish sign that our protests really are touching a nerve, and that they've agreed that they simply can't ignore them. So rather than address the problem by getting to grips with the climate crisis, they turn against the messengers. Classic stuff!

'Is this going to have a deterrent effect? A lot of people will continue in resistance come what may, even though it's horrible seeing friends and loved ones standing up against the state and being hit with these new legal powers, reinforced by a government that makes it very clear that it won't allow anybody to stand up to it. This is painful. And from a more pragmatic point of view, it's going to be harder to mobilise. It's a lot scarier to go out and take direct action if the consequences become ever more severe.'

Rosa

organisation deemed by them to fit into that category. This would include anyone who creates 'a permissive environment' for such organisations, providing financial support or even an invitation to them to speak!

Building on the work of Policy Exchange, Lord Walney's subsequent report, 'Protecting Our Democracy from Coercion', recommended that non-violent organisations like Just Stop Oil and Palestine Action should be completely banned, as terrorist groups, on the grounds, as he told the *Telegraph*, that such groups 'could promote unacceptably violent tactics in the future'. It's a shocking report, from a shocking right-wing extremist, who proudly acknowledges that he works for lobby groups representing fossil-fuel companies and arms manufacturers.

The UK is not alone in facing up to these threats to freedom. Back in March 2024, Liberties, the principal civil rights network in the EU, expressed concern that 'the rule of law is declining across the EU as governments continue to weaken legal and democratic checks and balances', noting in particular a sharp rise in restrictions on the right to protest. In the USA, right-wing organisations like the American Legislative Exchange Council (ALEC), alongside various oil and gas lobby groups, have drafted anti-protest laws that have been incorporated into several state legislatures. In April 2025, the International Centre for Not-for-Profit Law revealed that 41 new anti-protest bills across 22 states had been introduced since the start of the year. The *Guardian* highlighted one example: 'the Safe and Secure Transportation of American Energy Act would create a new federal felony offence that could apply to protests that "disrupt" planned or operational gas pipelines – which would be punishable by up to 20 years in prison or fines of up to $250,000 for individuals of $500,000 for organisations'.

United States NGOs are also having to deal with the growing number of SLAPPS (Strategic Lawsuits Against Public Participation) brought by large private-sector interests against civil-society organisations in exactly the same way as campaigners here in the UK are having to deal with more and more civil injunctions.

Some of the Just Stop Oil activists interviewed for this book have had to contend with both civil injunctions and criminal charges for the same action. Double Jeopardy barely begins to do justice to what this feels like. As George Monbiot explains: 'Both public authorities and corporations have been dropping injunctions on people who have protested, and, for that matter, those they believe might protest. Simply being named in an injunction exposes you to potentially massive financial penalties, as the named people – the defendants – typically have to pay the legal costs of the claimants.'

I often got the feeling, in these conversations, that the impact of all this, 'just one bloody thing after another', has perhaps been more onerous than my interviewees were letting on. And it's not difficult to imagine that this cumulative weighting of the scales of justice – new legislation interpreted more harshly, more restrictions on defendants in court, harsher sentences, more harassment, more surveillance, with police often making unannounced raids on activists' homes and lack of access to legal aid – was having a significant deterrent effect on Just Stop Oil campaigners before the organisation's decision in March 2025 to cease all campaigning activities.

There's a moment in every young Just Stop Oil activist's personal journey when they commit to the likelihood, or even the inevitability, of being arrested, prosecuted, and then – very likely, given the new laws – spending time in prison. I found it intensely moving asking each of them about that 'arrestable moment': how it happened; where it happened – on their own or with others; and whether they had really grasped the significance of that decision at the time that they took it.

On 27 September 2024, just hours after Anna Holland and Phoebe Plummer had received harsh sentences for damaging the frame of Van Gogh's *Sunflowers* painting in the National Gallery, three Just Stop Oil protesters repeated the action, 'souping' two Van Gogh paintings in the Gallery. One of them, Phil Green, is twenty-four years old. She would have gone through a similar process of self-examination regarding that 'arrestable moment' and was clearly undeterred by what had happened to Phoebe and Anna, no doubt intending to send a message to the government and the police to that effect.

But it would be very wrong to underestimate the emotional toll of all this.

'From the inside, we can see the strain which is being put on the prison system – too many prisoners, not enough staff – with so many people who are here for petty crimes such as shoplifting, as well as many people who need help with addiction or mental health. Most people can't believe that we're in here for climate protest – but are generally enthused to see people standing up against the Government!

'Last month, a Winchester magistrate made the brave decision to resign, declaring Winchester Prison to be "dangerous" and "corrupt". She said that she could "not sleep at night" knowing that she was incarcerating people in a place that would endanger them.

'Currently, 11 prisons in England have been judged to be unsafe, but people are still held in them. Across the prison estate, people are frequently left without the right medication, are subject to 23-hour lockdowns, a total lack of access to outdoor space for days and days, the rats, the broken windows, I could go on. Yet time and time again, studies show that sending people into unsafe prisons makes them more dangerous and therefore more likely to re-offend.'

Indigo

OLIVE BURNETT

I'm a twenty-two-year-old art and cultural theory student from Leeds, and I've been heavily involved in non-violent civil-resistance campaigns for more than two years.

INVOLVEMENT

My first involvement with the climate movement was a small uni group in Leeds doing an occupation on campus, then an Extinction Rebellion march with tens of thousands of people in London. As soon as I went to a Just Stop Oil talk, and then went slow marching for the first time, I felt in my gut that this was the place I should be. I had never heard people talk so clearly about the reality of our situation, and so openly about the seriousness of action needed to tackle it.

I have taken action to the point of arrest three times with Just Stop Oil, all slow marching in 2023. Then once in 2024 with Youth Demand, a newer sister campaign of Just Stop Oil, for painting '180,000 KILLED' – the total number of murdered Palestinians predicted from the ongoing genocide – in front of The Cenotaph war memorial in London.

MOTIVATION

I was desperate to get involved early on, but the weight of the realisation of climate collapse, and what that meant for all of us, felt too crushing for me to be able to cope with at that stage. I was frightened that if I tried to face my feelings about it, I wouldn't be able to continue with life on this insane planet. In a way, I was both wrong and right. Once I started, I was unable to continue with the life I had been living, but something new took its place and, although it's often stressful and frightening, it's also made me feel for once I'm actually really living. How often do people get to live lives where their actions truly align with their morals, and where they are surrounded by kind, brave people? I feel really grateful for that.

QUOTATION
A quote I often think of that helps me when I feel like panicking, is something another supporter of Just Stop Oil said to me: 'We're all making it up as we go along! We have to, because we're doing something radically new.' It regrounds me in the fact that this is all an experiment, and no one has the perfect answer, no matter how much we learn and strategise, but we have to try anyway and hope at the very least that those who come after us can learn and take courage from our work.

RESOURCES
The main resource I recommend to others thinking about moving into resistance is Roger Hallam's podcast Designing the Revolution – though be warned that the audio quality for much of it is terrible, but only because he recorded bits from prison! Also the documentary Bringing Down a Dictator, written and directed by Steve York, about Otpor!, the Serbian civil-resistance movement at the turn of the century, is one of my all-time favourites.

WHAT LIES AHEAD
I find the question of what success would look like a difficult one when so much collapse and suffering is already guaranteed. The truth is that real success looks like international action taken ten, twenty or forty years ago. However, there is still much we can do to limit the damage and there is still a choice in how we respond to the collapse.

For me, success is a total overhaul of how this country is governed: no more detached, profit-driven politicians working in a system that is designed to serve the mega-rich. It's ridiculous they are still perceived as having any legitimacy to make decisions that mean life or death for the rest of us. That is what all this work is about. It's not just the climate crisis, but the fact that it was possible for this to happen at all that has to be addressed.

OLIVER CLEGG

I was born in Stevenage, grew up near Cambridge, and now live in Manchester. I went to the University of Manchester to study biochemistry but quickly switched to plant science. After a year in industry, working in science communication, I'm now in the final year of my degree. I am an atrocious actor, but that hasn't prevented me from starring in seven amateur pantomimes. I'm also a massive nerd, especially when it comes to science fiction and Lego.

INVOLVEMENT

When I was a sixth-former, I went to my first XR protest out of pure curiosity. My perception of XR at that time was pretty negative: they were disruptive at best, terrorists at worst. And so I sat in the middle of a roundabout with these crazed eco-fanatics drinking tea and eating vegan brownies. And I thought, 'These terrorists don't seem so bad after all!' I went to Glasgow to protest at COP26 with Manchester University XR and a month or so later we blocked the road leading to an Amazon distribution centre on Black Friday. After that protest I was recruited into JSO.

After a run of particularly unsuccessful recruitment talks, the Manchester JSO student group adopted a more spectacular way of spreading the message: running onto Premier League football pitches. After JSO's first major protest, at the 2022 BAFTAs, I was arrested for the first time whilst running across the Spurs pitch. Then, little more than a week later, JSO's sustained campaign of shutting down oil terminals began. I was arrested four times in about a month.

For about a year I served as a spokesperson for JSO and became more involved with the LGBT group within JSO; with six other queer supporters of JSO, I blocked a road around Trafalgar Square in October 2022. The following summer I was one of the seven who sat in front of the Coca-Cola float at Pride in London.

After about a year away from arrestable activism, I took part in a Youth Demand protest, pretending to shit in the private lake at Rishi Sunak's Yorkshire mansion. The stunt was a visceral demonstration of my disdain for the disgusting state the Tories have left the country in.

MOTIVATION

Holding onto my motivation became harder as the police and government cracked down. I was stuck, reconciling my desire to make change, with my fundamental cowardice!

I see myself as someone who fights against injustice. To sound all pompous, I simply wouldn't be the person I want to be if I weren't an activist. And I'm very aware that, had it not been for activists of the past, especially the relatively recent queer liberation campaigns, my life would be markedly worse. I want to be a part of that continuing story of people fighting for better.

INSPIRATION

Peter Tatchell. No hesitation: Peter is undoubtedly my biggest inspiration. He was a key member of the Gay Liberation Front, and helped organise the UK's first Pride. He went on to co-found OutRage!, one of the UK's most significant LGBT direct action groups. Over the course of his life he has witnessed and created astonishing advances in queer rights. Peter is utterly relentless, unflinching and unstoppable.

IN NATURE

At heart, I'm a city boy. I'm embarrassingly unfamiliar with unspoiled nature. However, I don't believe that nature exists in a vacuum, separated from the communities who respectfully visit, looking for inspiration and leisure. So, I choose Knutsford Vale Park, a short walk from my house. When I wander off the paths, I can pretend that I'm alone in some sort of wilderness, and when I return to the path I am reminded that I am living amongst a density of brilliant people.

QUOTATION
'Get out on the streets or you're going to fucking die.' Larry Kramer

RESOURCES
Years and Years, the Russell T. Davies TV show, is probably the piece of media that pushed me towards activism. It features an activist character, fantastically played by Jessica Hynes, and throughout the series, as characters are more directly impacted by Britain turning to fascism, they all become embroiled in anti-government struggles.

When I saw the exit poll come through for the last election, with the projected number of Reform UK seats, Years and Years was all that we could talk about. The parallels were all too clear.

WHAT LIES AHEAD?
My hope has always been that, with JSO having successfully forced the government to stop licensing new fossil-fuel projects, we will inspire other climate activists to adopt similar tactics.

We will win. The worst effects of the climate crisis can still be averted. I believe that with all my heart, even if my brain doubts it. We won't go quietly: humans are persistent buggers! The community of resistance that we have formed will be able to weather whatever comes next, even if what comes next is fascism and famine.

(See p. 87 for Oliver's wry insight into being an activist.)

9: THE EMOTIONAL BURDEN

I PUT THE following question to the Just Stop Oil activists I interviewed for this book: 'How do you cope, personally, knowing what you do about climate change and the devastation that lies ahead?' These are some of the responses I received:

- 'I've learnt to live with it – well, sort of – but lots of crazy emotions. Up and down! Like normal people?'
- 'I hate being asked that question. I mean, what do you expect me to say?'
- 'I wouldn't be here if I wasn't shit scared about the future. I have been since I was fifteen.'"
- 'Fear – of course. Resentment – of course. Loss – sort of, though I've seen so little of the world that it's sort of "one degree removed".'
- 'Lonely – unless I'm with my Just Stop Oil group.'
- 'Sometimes I don't know how to get out of my Just Stop Oil bubble – how to be "normal", whatever that means!'
- 'Look, it's hard. Some of the things that are said about us, about our motives, are really painful. We all try and blank that out, but …'
- 'You probably think of us as a miserable bunch of eco-doomsters – but we still have a lot of fun along the way! Really!'
- 'I don't know – and I don't really talk much about it. Actually, I think we're all struggling. But is that surprising?'
- 'I'm fine, sort of, but I wish I had a better sense of where it ends – where all the rest of life starts – you know, work, jobs, partners, fun!'

- 'I sometimes wish I didn't know what I know, but there's no such thing as "unknowing" all this. Sorry, that really does sound like bullshit!'
- 'I know it's going to be bad, but what scares me is that perhaps I don't know *how* bad? And, honestly, hope is hard, with all that going on.'
- 'What I really hate is when people – friends even! – ask me where I am on some kind of superficial optimism-pessimism scale! What does that really mean?'
- 'Deep down, I'm grieving. But sometimes, that just feels like giving up …'
- 'Does it matter? It's "on us", right now, whatever it takes … and that's really not me being brave, it's just what it is.'

For me, this part of the interviews was by far the hardest. To be honest, I'm not sure how honest they were being. I don't know what the equivalent of a 'stiff upper lip' is for a young Just Stop Oil activist, but I came across it quite often. I realise that it takes more than a one-off interview to establish that kind of trust with people putting themselves in such vulnerable and emotionally turbulent positions. Even so, within those bounds, the openness and quiet determination were still mighty impressive.

A big part of the overall problem here – why we're still so bad at narrowing the gap between the science and the politics – is the relative ignorance of most politicians and media commentators. We essentially rely on huge numbers of decision-makers who really don't have a clue that climate change is not like every other 'issue', who know nothing about climate tipping points and the difference between reversible and irreversible changes. These are people who find it so much easier to read a balance sheet than to interpret data points and trends regarding the climate, and whose eyes metaphorically glaze over when invited to take full account of this existential risk to the future of humankind.

9: THE EMOTIONAL BURDEN

I guess this was part of the thinking behind Angus Rose's decision in April 2022 to go on hunger strike to pressure the Energy Minister at the time, Greg Hands, to organise a full briefing for MPs on the state of the climate. The Minister refused. Thirty-seven days later, Angus Rose had lost 17 kg and was getting very weak. At which point, a compromise was agreed when Green Party MP Caroline Lucas arranged for a briefing to be given by Sir Patrick Vallance, the Government's Chief Scientific Adviser, through the All-Party Parliamentary Group on Climate Change. Sadly, attendance by MPs was very limited.

In interviews at the time, Angus Rose made the telling point that it would be almost impossible for MPs to connect more deeply with their feelings without that basic level of understanding. And I really go along with him on that.

SYMPATHY FOR THE BYSTANDERS

I didn't ask my interviewees specifically about Angus Rose. But I was very curious to get a proper sense of their own emotional responses to what is happening, given that they are so much better informed than the vast majority of people today. And I pretty much got it all wrong in terms of what I was anticipating, which was an understandably unforgiving mix of frustration, inconsolable grief and a lot of anger – even rage perhaps. These emotions were indeed all in the mix for them, but not in the way that I had imagined.

For instance, speaking personally, I simply don't get why so few young people today share their determination to get stuck in to radical climate campaigning to try to make a difference, especially those at university with them. It is, after all, incontrovertibly clear that it's their generation that will be worst affected by contemporary society's wholly insufficient response to the Climate Emergency.

They turned out to be so much more understanding of this than I could ever be! 'It's really not the same thing being a student today as it

'There's a lot of what some people call "anticipatory grief", when you think ahead to the likely impact of climate breakdown on hundreds of millions of people, and on the rest of life on Earth. For a lot of us, I think that grief is more powerful than the anger that is always somewhere in the emotional mix.

'Personally, I don't have that much anger. I don't even feel like I have that much of an emotional response, bizarrely. I know that's a huge driver for some people, but for me it's just more black and white: "OK, I've got two options: either I do nothing, and let things go from bad to worse; or I try to do something, and hope it gets better." It's as simple as that. It's more a way of wanting to prevent more suffering – something which I feel deep in my heart.'

Rosa

'I sometimes compare myself with other activists and think perhaps I'm a bit less connected, even a bit "desensitised" by all of this. I have to remind myself exactly why it's worth putting myself through it – and sometimes, I just want to step back from it all, live a life that might be considered a bit more normal! I have a partner and it can get really exhausting thinking through the implications of everything, having that dread about the future. This can have such an impact on friends and family.

'We've had so many conversations in our kitchen, sharing how exhausting it can be, and getting fed up with the fact that things never seem to change much. I'm hugely inspired by people like Cressie Gethin. She's just the bravest person I know. Her moral courage and honesty shine through all the time, and I know that makes a big difference to others in the movement.

'Being in civil resistance is the most straightforward and healthiest way of coping with the existential threats we face.'

George

might have been back in the 70s and 80s' – even their put-downs were astonishingly polite! And it's not. Student fees, the future burden of debt, the cost-of-living crisis, zero-hours vacation jobs, insanely inflated rents for student accommodation and the wholesale depoliticisation of university life all make the experience of university very different from how it used to be. There are more than enough factors to deter students from getting involved.

Beyond that, they also appeared very sympathetic to their friends, who, aware of their activism and its consequences, didn't even want to get involved in climate conversations in the pub. This was one response:

> It does frustrate me, of course it does. But I guess they see what my life looks like and even though they know, instinctively, that the climate crisis is really bad, they absolutely don't want to have to confront it in the same way that I have. Let alone what it would mean for them and their hopes, dashed job prospects and so on. Look, I get it. And sometimes I find myself wishing that I was still in their shoes – with a lot less knowledge!

But not one of the protagonists in this book succeeded in missing that particular bullet, in all sorts of ways: through different experiences and influences, with knowledge building either gradually, over a few years, or rather more disruptively over a shorter period of time. Not one, by the way, pointed to what might be described as 'an epiphany', a one-off moment where what was formerly switched off was suddenly and irreversibly switched on. One way or another, that knowledge about climate change became an intrinsic part of them. And the implications of that knowledge have informed everything they have done from that time on.

EMMA DE SARAM

The system is working exactly how it's meant to: oil and gas companies profit from ecological destruction, while those who dare challenge the status quo are punished.

This is also a system that was certainly not designed for neurodiverse people who see the world differently. Yet within this very system, the intersection between neurodiversity and climate activism reveals both immense challenges and the potential for transformative change.

Like many neurodivergent people, especially women, I learned to mask my symptoms for years without a diagnosis. I spent years pushing through. In many ways, the climate justice movement became my arena of both action and survival. I threw myself into two years at the forefront of student activism: campaigning against a Shell sponsorship deal, demanding affordable food for students, resisting attempts to silence students who participated in Palestinian solidarity activism, and spending my time off campus on the streets taking part in Just Stop Oil actions. All this was fuelled by both fear of irreversible climate tipping points and my hope for something better.

But all this took a toll. Being the only person to speak out in a room of 30 university bureaucrats about the injustice of accepting money from Shell (the company that has caused destruction on a scale akin to that of a petrochemical empire), the exhaustion and anxiety of being 'that crazy activist', had to be pushed aside until breaking point, which manifested for me in developing a chronic illness. Going on to news outlets to speak what I thought of as common sense, being on the receiving end of endless misogynistic comments online, receiving constant trolling messages, would have pushed anyone to breaking point.

Only recently have I started to better understand my neurodivergence, and how it intersects with climate activism. Both battles are waged against systems that profit from harm, that exploit difference and marginalise those who resist. The struggle for climate justice and neurodiversity are inextricably linked, both products of an oppressive system built on ableism and ecological destruction.

Neurodiversity within climate activism is not just about overcoming personal challenges like mine; it's an opportunity to think outside rigid frameworks that dominate politics and society. But it shouldn't only be on neurodiverse people to challenge the system because our brains are designed differently.

Just as the climate crisis disproportionately affects marginalised communities, neurodivergent people can experience the compounded impacts of ableism, patriarchy and racism. My own experience of navigating neurodivergence is deeply intertwined with my personal experiences of patriarchy and misogyny. And while I obviously cannot speak for every neurodivergent person, it's clear that the systems that devalue our experiences are the same systems that perpetuate climate breakdown.

There's potential now to re-imagine climate solutions that transcend capitalist logics and foster systems that value well-being over profit. The current climate discourse centres Western capitalist solutions, erasing the wisdom of indigenous knowledge, or co-opting this knowledge and repackaging it to be profitable. Similarly, neurodivergent perspectives have been dismissed by a dominant narrative privileging 'normal' ways of living.

Creating a just and liveable future demands more than business as usual – we must re-imagine the systems that brought us to this crisis. True climate justice values neurodiverse thinkers and marginalised voices, for they are already envisioning a world beyond the one that fails them. And fails us all.

'I don't regret anything I've done, but it sometimes makes me quite sad – though that's not really a strong enough word – that I'm not able to live the more normal life of someone in their early twenties. I do sometimes grieve for the life I could be living. For instance, I have a twin brother, and he's just graduating and will soon get a job – like most of my fellow students. I sometimes look at my peers and wish that I was doing what they're doing.

'I'm able to do this because I'm not alone. If you're just waking up to the scale of the climate crisis, alone in your bedroom, without any infrastructure of support around you, it's truly awful. I'm only able to confront this crisis and take action because of the community around me.'

Ella

'I'm still enrolled for my degree, but I don't plan on finishing it. That's not been an easy choice, but I can no longer see the point of finishing it – getting into research, following that kind of career path – after another ten years of research, what then? Societal collapse? This has not been an easy choice, and I know it's been really difficult for my parents. For my dad, a degree has always been something to fall back on, come what may, something that you've always got in reserve.

'The truth is that it's a bit of a privilege to be living through this, to have the opportunity to get involved in something so critical, to meet so many amazing people. There's something about being involved in a cause that really can change the world around you. I'm just grateful to be a part of it.'

Daniel K.

9: THE EMOTIONAL BURDEN

CLIMATE ANXIETY

This brings us to the phenomenon of 'eco-angst' – a spectrum of emotional and psychological responses to the combined climate and nature emergencies, about which huge amounts have been written over the last couple of decades. The academic literature on this is already voluminous, and outlets like the *Mail Online* delight in the opportunities that eco-angst gives them to excoriate today's 'snowflake generation' and its 'spoilt brats'! Paul Marshall (hedge fund billionaire with massive investments in fossil fuels, owner of the *Spectator*, GB News, and prospective owner of the *Telegraph*) endlessly trots out his would-be witticism that they're all suffering from 'climate derangement syndrome'.

But whatever the *Daily Mail* may say, there is no doubt that eco-angst is a real phenomenon. There's even a fancy name for it – 'solastalgia', or a feeling of distress associated with the loss of something close to one's heart in the environment. According to a large survey done by *The Lancet Planetary Health* back in 2020, 84 per cent of young people in the UK describe themselves as worried about climate change, and 59 per cent reported that it has a negative effect on their daily lives.

At the very least, the kind of 'unravelling' of the Earth's natural systems and the simultaneous disruption of communities and society as a whole will inevitably impose psychological burdens on significant numbers of people. In writing *Hope in Hell*, I touched on the phenomenon of what some are already calling 'pre-traumatic stress disorder', where even those not yet directly impacted by climate change are affected by the prospect that they and their families and loved ones might be in the future.

Beyond that, a growing number of scientists involved in the fields of anxiety, depression, dementia and neurodegenerative diseases are beginning to share insights between them, teasing out all the different ways in which climate-related impacts both chronic (such as rising temperatures) and disasters (such as floods, storms and heatwaves) are affecting people's health, physiologically as well as psychologically.

'Just Stop Oil uses quite a few spokespeople, and we all have different areas of expertise. I've kind of switched off my emotions to make it possible to go on doing this for Just Stop Oil. For one thing, I have to stay on top of the climate science, following all the climate disasters, in order to do the job properly. It just isn't right to try to sugar-coat the consequences of all this, and the extent of the devastation ahead. But if I was constantly connecting with the emotional impact of all that, it would get impossible. I'd end up crying on camera, showing my true feelings, and I'm not sure that would really work!

'Psychologically, it's incredibly tough to be in that place, and the reality is that the more closely people are connected to those emotions, the more likely they are to burn out. I've had to address that head on, to avoid burning out, but that also takes its toll. As does feeling angry all the time about what's going on – that can really eat one up inside – and becomes another cause of people burning out. Yes, I'm furious about the fossil-fuel companies and governments, but my main emotion is sadness rather than anger.

'Colleagues who have been doing this activism the longest don't act any longer out of big emotions – they act out of a sense that it is simply the right thing to be doing. Or even the *only* thing to be doing. It does make one more resilient, but there are obviously big risks in terms of personal well-being.'

Alex

'Every day we're seeing that more and more young people are suffering from anxiety – I often used to feel overwhelmed by climate grief myself. And that's on top of everything else, after so much has been taken away from us in terms of expectations about jobs, housing and basic security. Added to this is the trauma of seeing other children and young people starved and murdered every day in the news.'

Anna

That additional burden may well come to weigh much more heavily on health professionals and affected citizens in the future. Right now, even the relatively limited knowledge about climate that young people have is inevitably having some impact. Most young people have been exposed from early on to the basics of climate change, and other impacts on the environment, arising from our industrial, consumption-driven way of life. These topics are well represented in today's school curriculum, though the extent to which they are prioritised depends a lot on individual schools. Even the most anodyne of climate forecasts has the potential to make young people fearful of the future. For some, as their knowledge deepens, that fear can become debilitating.

All those I interviewed are familiar with these emotional and psychological realities. They all have friends, colleagues or siblings who have experienced anxiety, fear, burnout, dread, helplessness – even hopelessness – as a consequence of worrying about climate change. And they've all wrestled with these demons themselves. For all of them, taking action and entering into civil disobedience is cited as the most effective way of keeping those fears at bay, of protecting themselves from despair. I would paraphrase it as follows:

> It's impossible not to be fearful of what the future holds for our generation – and for the whole of humankind – if we carry on with our current way of life, with emissions of greenhouse gases continuing to rise every year, with climate disasters multiplying all around us. Not to be fearful is not to be alive, to be thinking, feeling, connecting to others. But we also know this could all change, and could change very quickly, and it's still not too late for such a change to prevent the very worst scenarios that we read about. So, every day that I'm able to devote myself to helping that change to happen, through direct action and civil disobedience, the less fearful I am.

'It's been very interesting for me having this enforced reflection time here on remand. To start with, I really didn't cope with that very well at all, just in terms of the contrast between the "go, go, go" intensity as a Just Stop Oil organiser, and then everything suddenly closing down around me. Things got easier when I was able to develop some kind of routine.

'It's a good opportunity to think about what it means to be a "political prisoner", especially in a historical context. People talk about us in terms of the "sacrifice" that they think we're making, and although that's obviously true at a certain level – not leading an ordinary life, not living with one's partner, unable to look forward to new job opportunities – it's not a word that I use myself. I still feel compelled to do what I know needs to be done, to be in civil resistance, and I undoubtedly get a sense of fullness and joy knowing that I'm able to contribute in that kind of way, however big or small that may be. That's definitely a sort of blessing.'

Indigo

'People can't stop living just because of the climate crisis. They can't stop finding joy in life. For me, that has a lot to do with the environment, as I've always loved nature from a very young age. I find a lot of peace being outdoors, with strong connections to the land and nature around me, so I guess it's not a surprise that I'm now doing an outdoor education degree course and, hopefully, will soon end up as a qualified mountain guide.

Scotland used to be like a rainforest, and now every single hill is a desert. It's not easy trying to be a mountain leader in something as badly damaged as that, but then you start thinking about all the people who are working away rewilding those places, helping nature to flourish again. And I find that very exciting. That's my way of keeping going, even as we have to keep on dealing with the climate crisis.'

Eilidh

When asked if they would feel the same doing more conventional campaigning, without the commitment to direct action and the possibility of arrest and prison, the answer was invariably no – not least as they've almost all been involved in that kind of mainstream campaigning before committing to civil disobedience. And when asked if they would feel the same even if they had come to the conclusion that it was already too late to prevent the very worst scenarios that they read about, the answer was usually yes – on the grounds that there is always going to be something to fight for, some way of stopping things getting even worse.

There is an inner 'emotional logic' to all this that I found impressive – and moving. In that regard, it's impossible to exaggerate the critical importance of the role that Just Stop Oil itself plays in supporting a community of like-minded people, all of whom are refusing to turn away from the physical reality of accelerating climate change, refusing to continue to act as bystanders, and all of whom are persuaded that it is only civil resistance to the status quo that entitles one to hang on to authentic hopefulness. All of them talked so warmly of the constant support they receive from their fellow activists and from the organisation itself, including the Climate Action Support Pathway, of the relief of knowing that they are understood without endlessly having to justify or explain, and of unspoken solidarity, providing a somewhat more reassuring backdrop to their lives as activists.

THE LANDSCAPE OF GRIEF

But no amount of solidarity can immunise people against the grief caused by the climate crisis. This is now a much more familiar aspect of the media coverage of climate-induced disasters: farmers staring out over flooded fields, seeing years of care and hard work destroyed in days, or even hours; homeowners standing in the charred remains of all that they once owned, not just bricks and mortar, but memories of a lifetime; parents or family grieving the death of loved ones caught up in

'Resilience means being able to keep some sort of balance between hope and despair. There are two things that are particularly helpful for me. First, it's that agency thing, knowing that I'm doing as much as I possibly can, and will be able to look my future self, my future family, in the eye and say I did everything I could, putting my body on the line. Before that, I'd often feel guilty about things, knowing that I was inevitably part of the problem – and correspondingly powerless. Second, up until a few years ago, I naïvely hoped that everything would be OK, and would somehow get sorted out! Well, everything won't be OK, and my hope now lies in doing what we have to do in the near future to avoid the worst.

'I don't necessarily like to admit it, but I know I'm motivated by a sort of despair as I see the things that I love in the natural world disappearing. I'm fortunate in that I'm able to channel that into relatively safe actions. I often end up approaching climate activism with an almost meditative perspective, just trying to do things in a very calm way. But the truth is that I'm absolutely heartbroken, and therefore absolutely furious that the people in power in politics and in business just allow this to continue.'

Ollie

'I'm a pretty cheerful person most the time, and definitely not "despairing" by natural disposition! Of course, thinking about the climate crisis makes me feel stressed, anxious or even angry sometimes, but that's not the same thing as despair. I can see there might be some kind of "soft denial" going on here, as I struggle to imagine just how bad it could really get in the future. I guess this is a form of self-protection – I certainly don't sit there thinking systematically about what will happen when we have a couple of metres of sea-level rise, and I'm there caught in the flood-waters trying to escape!'

Oliver

extreme storms and weather events; the unbearable suffering of people whose homes are flooded, sometimes on a regular basis.

This is the landscape of grief with which more and more of us are now being confronted – and it's felt very powerfully by Just Stop Oil campaigners. There have always been 'natural disasters' throughout human history, inflicting loss and sadness on people no less intense than what we're becoming aware of today as a direct consequence of the climate crisis. Human suffering as a result of a lethal storm, an uncontrollable wildfire or a devastating mudslide is the same after a natural disaster as after a climate-induced disaster, or a combination of the two. But our growing awareness of the degree to which our species is now responsible for the increased frequency and intensity and the worsening impacts of such extreme weather events, indirectly changes the very nature of the grief we feel.

As climatologists endlessly remind us, claiming that any one of these events is a direct consequence of climate change remains problematic. But the increasingly authoritative science of attribution – determining the contribution of climate change to specific extreme weather events – is making it much easier to get our heads around the scale of the changes that are taking place. An extraordinary study published by *Carbon Brief*, in November 2024, drawing on a new database made up of hundreds of different studies analysing the role of extreme weather events, or 'un-natural disasters', revealed that at least '550 heatwaves, floods, storms, droughts and wildfires have been made significantly more severe and more frequent by global heating – including at least twenty-four 'previously impossible' heatwaves that have struck communities across the planet, with scientific analyses showing that they would have had virtually zero chance of happening without the extra heat trapped by fossil-fuel emissions.' ('Climate Crisis to Blame for Dozens of "Impossible" Heatwaves', Damian Carrington, *Guardian*, 18 November 2024.)

Hearing things like this, the understandable next step for young campaigners today is to point out to anyone who'll listen that this is

'My work with Just Stop Oil has been the most motivating that I've ever done. I have to be careful about not getting arrested, on account of my visa, so I do a lot of welcome talks, facilitating Zoom meetings and data management. And it's the most incredible community that I've ever been part of, although one thing I've struggled with is maintaining my friendships outside of Just Stop Oil. A lot of those friends are very focused on careers, and on the kinds of things that I just don't find valuable any longer, and while they try to be supportive and understand what I'm doing, I'm not sure how deep that goes.

'We all recognise that the climate crisis takes away whatever dreams we may have of a better world. That grief is present in all of us, and it's very painful. Personally, I take a lot of comfort in the work that I'm able to do with Just Stop Oil, but I'm conscious of the need to reground myself all the time, to remind myself of what really matters. I hope this doesn't sound too morbid, but it's almost as if we have a duty to appreciate what we've still got, while we still can.'

Avery

'I feel I've been to hell and back thinking through the implications of the climate crisis, and that's what makes me willing to do whatever it takes to avoid that nightmarish outcome. I sometimes wondered if I would be able to come back from all that, from that horror story, and carry on with the work – this kind of emotional intensity can be almost impossible to live with, especially when one takes into account the possible rise of fascism and violence as part of the consequence of what is happening right now.

'And I can sometimes get very angry, when I hear about Shell's latest strategy, or something like that, knowing that this will just blow through our carbon budgets. But that's not a constant undercurrent for me, and it's certainly not what motivates me in the work I do.'

Eddie

all happening – right now – with an average temperature increase of no more than 1.3°C since the start of the Industrial Revolution. 'So, what do you suppose it's going to be like, for people all around the world, at 2°C? Or 2.5°C? Or 3°C?' As concerned citizens, not even as climate activists or climate scientists, how can we not extrapolate from what the landscape of grief looks like at 1.3°C to what it will look like at 1.5°C, which is now locked in, and beyond? The burden of anticipatory grief grows heavier for every wasted year that passes.

Regarding the scientific community, I've followed the inspirational campaigns of Scientist Rebellion since 2020, as they've taken action in the UK, Germany, Spain and the US. Their rationale is crystal clear:

> Scientists have spent decades writing papers, advising governments, briefing the press; all have failed. What is the point in documenting in ever greater detail the catastrophe we face, if we are not willing to do anything about it? Academics are perfectly placed to wage a rebellion; we must do what we can to halt the greatest destruction in human history. Non-violent civil resistance has proven to be one of the most effective tools to catalyse change, and has been a major driver behind historic victories for decolonisation, labour rights, women's rights, LGBTQ+ rights and civil rights.

But the number of scientists involved is still relatively low, and I sometimes find myself asking why so few scientists – many of whom are themselves parents or grandparents, unequivocally aware as they surely must be of what a world 2.5°C warmer than our world today will look like – have become actively involved in civil resistance? For understandable reasons, no doubt: most have jobs, families and financial responsibilities and professional obligations. It's a big ask to expect them to put all that at risk.

'I used to be a lot more motivated by frustration than I am now, and that really impacted on my family relationships and friendships. I just ended up blaming the wrong people, or, if not blaming them directly, getting increasingly frustrated at people not making the connections – in terms of what we're eating, for example. I know that was completely unreasonable, because I too once enjoyed a typically consumption-focused lifestyle! But sometimes it feels like we're walking around in a consumerist hell-hole, with everybody just averting their eyes.

'I'm conscious these days of the need for a better balance in my life, and I'm happy to be able to support my mum in running a repair café that we set up a few years ago. That kind of personal responsibility is always going to be a crucial part of the bigger picture.'

Emma

'Most of the time, I feel I just have to find a way through, not allowing myself to be too burdened with all the pain and oppression going on around us. I still have to read a lot about what's going on in the world, and the possibility of traumatic breakdown and the collapse of the food system. But to go on doing the work, I need some kind of emotional insulation. If you have too much of an open heart, you'll just be eaten up and spat out.

'I can't say that anybody ever taught us how to do this kind of campaigning, but you do need to have a good level of resilience. It's all too easy to get blown around in the wind. I meditate every day, and I've read a lot about Buddhism, and get a lot of support here. Beyond that, Just Stop Oil has a really healthy culture, even though we're aware all the time of all the shit going on out there, putting an incredible burden on us. But we have a very solid community and, to be honest, just doing the work, being on the front line, doesn't leave me much emotional or mental space to be worrying too much about that side of things.'

Sam

9: THE EMOTIONAL BURDEN

GRIEF, LOSS AND ANGER

But that's exactly what young Just Stop Oil activists have chosen to do. I asked all of them to tell me what they feel they have lost because of the actions they've taken. Not one regretted the choices they'd made – or, at least, they certainly weren't telling me about it! But all feel, to varying degrees, that they've lost part of what it might be like to be 'a normal young person', unencumbered by climate responsibilities. Many have lost friends; relationships with parents and siblings have often been fraught, sometimes to breaking point. Some have given up their place at university, sceptical about the value of whatever degree they might otherwise have got in an increasingly disrupted world; others are continuing with their degree courses, insofar as they are permitted to do so by their universities.

Many have given up what might be described as 'the prospect of a conventional career path', with so much uncertainty ahead of them personally and for society – even though they derive some comfort from the likelihood that whatever contribution they can make in that disrupted world is unlikely to be any less valued than it is in today's world.

And all, in one way or another, have given up any default assumptions that we can automatically look forward to a better world for people in the future. There might still be a better world, but they certainly don't spend much time dreaming about what that world might look like. What is more, they implicitly accept that any variation of 'better' will have to be fought for as resolutely as they feel they're now fighting for the right to live in a world with a stable climate and thriving life-support systems. Their eyes are wide open to the inevitability that democracies the world over will be even more at risk tomorrow than they are today. 'That just comes with the territory.'

One might imagine that such feelings of loss would leave anyone affected with a deep sense of grievance. Or anger. Rage, even. You'd be wrong – at least as far as most of the twenty-six are concerned. Very few acknowledged, unprompted, that anger plays a big part in

'The climate crisis is obviously more emotionally resonant for me than it is for most people, but I know I'm not as deeply connected as some of my friends in Just Stop Oil. I think I would go completely crazy if I had to hold that prospect of intense human suffering at the front of my mind, day in, day out. So, there's a real paradox at work here: you have to be able to connect with that reality, to feel genuine empathy with all those whose lives will be destroyed by the climate breakdown, otherwise it really wouldn't be possible to take this work as seriously as it needs to be. At the same time, you have to be able to disconnect from all of that just to be able to function at a basic level and not feel crushingly depressed. It's a hard balance: how engaged do you have to be to want to be in a radical civil resistance campaign, and how disengaged do you need to be to be able to get the work done?!

'That all sounds very abstract, but it links deeply to your own personal emotions. I've had a lot of grief in my personal life, as have a lot of people, so I find myself connecting to what that climate grief is going to be like for people in the future at a very personal level.

'I've given a lot of talks on behalf of Just Stop Oil, and I'm aware that a lot of people at these talks are weighed down by some kind of eco-angst, fear or sadness. So, I just tell them that I used to feel like that, isolated and hopeless, disengaged from things going on around me, but that being involved in the movement has changed so much in my life. Instead of being paralysed by all those emotions – by fear and grief in particular – addressing the cause of all that becomes energising, connecting and really powerful.

Olive

their emotional response to the Climate Emergency. And even when they were prompted, by me, a very mixed picture emerged, depending on what they saw as 'the awareness gap' amongst different groups of people. Executives in the oil and gas industry emerged as the principal target of any anger, based on the fact that they've had greater access to the science of climate change, and its implications for the future, than any other single group of people in society today.

Interestingly, however, that anger did not seem to extend to people of their own age still enthusiastically taking jobs in the fossil-fuel industry, even though they too must surely be as aware as any young person today of what is going on.

There is certainly some anger when it comes to the dogged adherence of mainstream environmental organisations to conventional campaigning tactics – as touched on in Chapter 1. For many of those I interviewed, the Moderate Flank's 'theory of change' looks entirely threadbare, so dependent on 'magical thinking' about the role of new technologies and even, in some ways, so self-serving in helping to maintain the status quo that it is more of a 'flaccid flank' than a Moderate Flank. Those I referred to in Chapter 7 as 'the double-downers', sticking to the same tactics as before – raising awareness, lobbying MPs, organising petitions and marches – but just doing it all much more effectively – do not command much respect. Doing the same thing over and over again and expecting different results is indeed insanity.

But I detected no personal animus here; they see this is just one aspect of a broader generational issue, part and parcel of the rather low expectations so many young people have today of their parents and grandparents in acknowledging the shrinking legacy they'll be passing on to them. I return to this whole question of Intergenerational Justice in the next chapter.

When it comes to politicians, the dominant tone appears to be more 'weary disappointment' than burning rage:

'They're all trapped in the same system, just as much as anyone else.'
'There's just so much "fire-fighting" they're involved in.'

'Although I can talk about this quite rationally, the prospect of ending up in prison is not good, and the likelihood of being raided by the police means that you're sleeping with one eye open, and it's stressful not knowing what can happen with some of these legal processes, with stress levels basically burning your nerves even as you find yourself juggling with a range of different emotions: grief, at the prospect of devastation caused by climate breakdown; fear, inevitably; and quite a lot of anger, as you can see, knowing that we are now locked in to the death of hundreds of millions of people as a consequence of something which could have been avoided. I always make the point that I was born after the first big climate conference, which basically means that I've been systematically lied to my entire life!'

Chiara

'There are a lot of us who are always questioning whether what we're doing is enough, whether we should be sacrificing more, even as we end up having to persuade friends and family that we're not crazy as it is! I've realised that I'm much more resilient than I thought I would be, especially when the police raided my home last year. However, I've also come to realise that I'm quite fortunate in having layers of separation in my life. I have a job, in a climate-conscious business, that I really have to work at, and at the end of the day, I can still go home, message friends, play a video game and so on.

'It's not so easy for people who work full-time with Just Stop Oil, with hardly anything outside of their life that isn't about the cause. That's often not sustainable, and I've seen a few incredibly strong, brave people, with inspiring integrity, burn out completely with the emotional highs and lows of working against the state that does everything it can to crush them.'

Jacob

'So many things going wrong, and so many short-term priorities to be addressed.'

I have to admit that this astonished me. Try as I might – which isn't very hard these days! – I cannot find it in my heart to rationalise the vast majority of politicians' indifference and inertia, let alone to forgive it. And the fact that the wholly corrupt nexus of interest between politicians, the fossil-fuel industry and the media still goes on today, in so many different ways, still enrages me. It enrages me to the extent that I believe every single one of the young people interviewed for this book has been directly and systematically betrayed by today's politicians. Not knowingly, perhaps, in the past. But indisputably, right now, knowingly.

For the most part, the young people do not share that rage. I felt constantly rebuked by their more balanced and tolerant gaze.

OLLIE SWORDER

I'm twenty-one, a Masters student at the University of Oxford, studying to become a conservation researcher. I'm also a musician, singing in a choir, and spend much of the rest of my time learning about conservation or actively researching to make the world a better place – I'm researching beaver reintroductions to inform policy to support more reintroductions across the country. I see my life in science and academia and live that life through science and activism, in equal measure.

INVOLVEMENT
Extinction Rebellion (XR) – Marches 2019–22, without arrest; local group actions (Winchester) 2019–21; attempted, unsuccessfully, to start a student XR group in Oxford in 2021.
Just Stop Oil, 2023–24: banner drop at the uni boat races 2023; slow marches without arrest in the summer of 2023; Just Stop Oil students launched in late 2023; slow marches (arrested in November 2023, found not guilty.)
Youth Demand 2024: disrupted a Labour fundraising dinner before Labour was elected.

MOTIVATION
I have always been driven by being well informed on the science of climate and nature, by a sense of both grief at the loss of nature and by anger at the destruction of it by corporations and our governments. I have been taught by the top academics in the world at Oxford, and have had lecturers on the verge of tears describing the complete breakdown of the natural world and modern society. I am therefore driven to do everything I can to prevent this destruction.

This started as putting my mind to use, through science and communication, but has also become me putting my body on the line out of desperation, because nothing less appears to be able to shake up 'business as usual'.

There is also the grief that, if nothing changes, my younger brother and my future children will grow up in a world of suffering as food shortages create unrest, and as our own family and friends die from heatwaves or lose their livelihoods to flooding or as more pandemics ruin our lives. Knowing the stakes, knowing that, given current trajectories, I have already lost everything, I have nothing more to lose by putting my body on the line.

INSPIRATION
I would have said Greta Thunberg two or three years ago, but nowadays it's Daniel Knorr (see p. 99), who I am very glad to say has become a good friend of mine over the last couple of years. Their courage in the face of state repression, and their utter kindness and strong morality have had a great impact on the way I see the world.

IN NATURE
I feel at home simply in nature, but especially when I see or hear bats or red kites.

QUOTATION
'I suffer for the future of this planet, out of love.' That was the sentence that came to me in my police cell following my first arrest.

WHAT LIES AHEAD
In the best-case scenario, our government is moved by public pressure into setting more ambitious climate goals, and also massively increasing the action required to meet them. To me, that means something like setting a target, axing all the Tory oil licences like Rosebank, harsher taxation on fossil-fuel companies to pay for a massive scale-up of investment in green infrastructure, accompanied by wide-scale dissemination of accurate climate information.

(See p. 242 for Ollie's plans for writing a book.)

PAUL BELL

My name is Paul, but I usually go by Pasha. I'm twenty-four, doing a PhD in statistics at the University of Exeter, looking at extreme precipitation and climate impacts.

INVOLVEMENT
I got involved in climate activism in 2020 with Extinction Rebellion. I was arrested for the first time with ER at the August Rebellion in 2021, and then in March of 2022 for an action focused on Barclays Bank's fossil-fuel investments. I took action with Just Stop Oil with the oil terminal blockades in April 2022, and then continued to take action with Just Stop Oil throughout 2022. In total, I have been arrested nine times. In November 2022 I took action with Just Stop Oil by climbing a motorway gantry over the M25. I was remanded in prison for forty days, then released with an electronic tag. I was finally sentenced to prison for this action in August 2024. I am now on home detention curfew following my release on 10 September.

MOTIVATION
My motivation is that I think that resistance in this time of existential crisis is the right thing to do. Now that I have accepted the reality of climate collapse, and all the other aspects of the global death machine of capitalism and colonialism, opposing that death machine is the only option.

INSPIRATION
I am most inspired by the Suffragettes, especially by a woman called Kitty Marion, who ended up being force-fed 232 times, and bore that torture despite her work as an actress requiring her voice.

IN NATURE
I love birds. I recently saw a kingfisher, which was very wonderful.

QUOTATION
'Only do what you're comfortable doing, and then just a little bit more.' My friend said this to me, and it has always rung true.

RESOURCES
I recently watched Love in the Time of Revolution about the 2019–20 Hong Kong student protests, which I found very inspiring and moving. It shows what a mass street movement could really look and feel like.

WHAT LIES AHEAD?
I don't think any of this will end in a couple of years, though obviously that is not a reason for apathy. A better world starts when the media isn't owned by billionaires, when corporate lobbyists don't have influence over governments and when democracy is more direct and deliberative – for example, via Citizens' Assemblies.

'My parents were always very aware of what was going on in terms of climate change and ecosystem breakdown. My dad volunteers with the Avon Wildlife Trust at weekends, and both are deeply concerned. But they didn't really like talking to me about it! We watched a lot of documentaries, but then my dad started saying he found even David Attenborough just too depressing – I think he probably didn't want to face it himself and not burdening us with it gave him something of an out!

'Both my parents have now been arrested for their involvement in various slow marches – my dad was acquitted; my mum has still got a trial coming up. My grandmother also joined one of the marches. All this was very emotional – for all of us – especially the night before the action which we were doing together.

'I've now been arrested eleven times and still have two Crown Court trials ahead of me – one for the same gantries action that George Simonson (see p. 128) was found guilty of and sentenced to two years for. I know it sounds weird, but despite all that, I've got used to living with all these things hanging over me.'

Sam

'I have one very strong memory from when I was six or seven years old. A couple of speakers from an environmental organisation came to our primary school to talk about climate change. And one of them told us that we only had so many years before it would become impossible to reverse the damage we'd done. As she said this, she looked down at the ground because I don't think she could bear to say that in front of a whole sea of primary-school kids. I don't think she could look us in the eye. As I grew up figuring out who I was, things moved on. But there was always that latent memory that stayed with me.'

Jacob

10: INTERGENERATIONAL JUSTICE

'Only when it is dark enough can you see the stars.'
Martin Luther King

AS I EXPLAINED in Chapter 1, the idea of Intergenerational Justice is what gave me the idea of writing this book. Although it may sound a bit woolly, for me it's anything but. It's been hard-edged and highly consequential – not just in the sense of it being 'something important', but in having had major consequences for my politics, my career and for my sense of what it means to play out one's role in life as responsibly as possible.

I took over as Director of Friends of the Earth (FoE) in 1984, after ten years pursuing a dual-track life as a teacher and with the Green Party, or Ecology Party as it was then. FoE's positioning was very forward-looking, seeking to sort out problems in the present to avoid grim consequences in the future. We were given a massive boost by the publication of 'Our Common Future', usually known as the Brundtland Report, in 1987.

This report has little if any resonance in the broader sustainability movement today. It's pretty much dead and buried, a bit like Fritz Schumacher's *Small Is Beautiful*, albeit with a very impressive headstone. But 'Our Common Future' was much more than a report. It was the culmination of years of work undertaken by a high-level commission chaired by Gro Harlem Brundtland, the then Prime Minister of Norway. At its heart are two big ideas: that social justice and environmental sustainability are two sides of the same coin; and that we had to start bringing the interests of future generations to the centre of contemporary policy-making. In Gro Brundtland's own words:

'Even for someone like me, in the midst of it all, it's really hard to get one's head around the fact that we might actually be destroying the very foundations of life that made human civilisation possible, that we could have the equivalent of something like the plague wiping out entire continents, or the total destruction of human society, not through any other force of nature but ourselves. And what we set in motion cannot now be undone – the genie is out of the bottle, that's for sure.

'It sometimes seems a bit odd to be doing a PhD at a time like this! In some ways, I know it's just totally abstract nonsense, with a veneer of academic respectability. For me, it's just fun, doing puzzles and algebra, even though I know I shouldn't really be doing puzzles when our house is on fire! But when you're in a police cell, your world tends to narrow down quite a lot, and those puzzles became food for my soul at that time – it's a really helpful survival strategy to be able to stare at a blank wall and keep one's brain active!'

Chiara

'There's another part of this: a lot of people just don't feel any hope about the future, don't think that actions amount to anything, don't feel that kind of empowerment. I know how lucky I was that I met people who *did* have that hope and could inspire me to feel the same empowerment. That's what so many people are missing. And it's really sad, because when you look back at history, basically every single right we have, or every single thing we now take for granted, has been won by people having that hope and having that empowerment to go and do it for themselves, and really fight for it. It's sad that we've got to the point where people have almost forgotten about how rights work, and about rights not being given to you but having to be fought for.'

Eilidh

10: INTERGENERATIONAL JUSTICE

> If we do not succeed in putting our message of urgency through to today's parents and decision-makers, we risk undermining our children's fundamental rights to a healthy, life-enhancing environment ... we must translate our words into a language that can reach the hearts and minds of people young and old.

Bringing social justice and environmentalism together was a big deal back in the 1980s. At that time, the Environment Movement and social-justice organisations operated in almost completely separate spaces. Greenies couldn't understand why international development and aid organisations, for instance, didn't put concerns about the physical environment – not so much about climate change in those days, but about pollution, deforestation and contamination of water sources – at the heart of what they were campaigning for; and most poverty and justice organisations tended to think of environmentalists as 'the well-to-do protecting the nice-to-have', as one memorable line put it at the time.

The Brundtland Report undertook some banging together of heads at the highest level, essentially suggesting that both critically important social movements should rapidly evolve – given that each was completely dependent on the other for achieving mutually-inclusive objectives. Nearly forty years on, I'd say progress on this front has been mixed – as indeed I picked up in a lot of my interviews for this book. For instance, the links between the genocide in Gaza and the genocidal impact of climate breakdown on people in the future, which is self-evident to just about all Just Stop Oil and Youth Demand activists, is certainly not self-evident to the majority of environmentalists today – let alone to mainstream environmental organisations.

The single most enduring legacy of the Brundtland Report is its definition of sustainable development – 'development that meets the needs of the present without compromising the ability of future generations to meet their own needs'. But the report itself was not uncontroversial, with most environmentalists very uncomfortable that reconciliation

'I know I've been very lucky in that I've had the support of my parents all the way through this – both of them have science degrees and are really concerned about the climate crisis. I'm not sure I would be doing it in the way that I can today without that support. Emotionally, I know it's really hard for some of my colleagues who just don't have that kind of parental support.

'Obviously, it makes a huge difference to be surrounded by people of all ages who understand what's going on with the climate crisis. But it's also the least transactional community I've ever been part of – people are incredibly generous. It probably sounds a bit corny, but there is a real sense of service, and incredible empathy for other people. And what I hadn't expected was this intergenerational aspect, which means I'm working with people who I would normally never interact with on a day-to-day basis.'

Avery

'All I've ever really wanted is to grow up, be a writer, and have a big family! I come from a big family, so I've always wanted that. I'm twenty-two now, and I already know that I won't have children – I wouldn't bring children into this world. So, at twenty-two, at the very beginning of my life, and knowing that this part of my life is already over, is one of the worst feelings. It's commonly known that one of the worst things a parent can go through is the loss of their child. It's the one thing that any parent wishes to avoid with every fibre of their being – that they won't have to bury their own child. I'm still a child myself, in many respects, and yet I'm already grieving the children I've yet to meet. I still have my children's names picked out. I still imagine a future where I have children, and a house, and a family. And then I'm reminded, every time I look at the news, every time I go online, every time I look outside my window, that those children that I imagine I would love so deeply, I'll never know them. They'll never know me.'

Anna

between the interests of present and future generations was seen to be entirely dependent on continuing high levels of conventional economic growth. Even then, this kind of no-holds-barred economic growth was recognised as the principal driver of environmental damage.

Despite that, 'Our Common Future' gave everyone a genuinely 'big idea' that could be amplified and brought to bear on the unreconstructed politics of the day. And so it proved to be. By the time of the Earth Summit in Rio de Janeiro in 1992, five years after publication of the Brundtland Report, there was a significant groundswell of support for 'doing economic development differently'. Although the Earth Summit is best known for the two international treaties which resulted – the UN Framework Convention on Climate Change and the Convention on Biological Diversity – equally important outputs included the Rio Declaration, still as evocative a 'mission statement for humankind' as you'll find anywhere, and Agenda 21, a sprawling compendium of policy proposals and priorities.

SUSTAINABLE DEVELOPMENT

In many countries, the impact of Agenda 21 was felt as much at the local level as nationally. Here in the UK, for instance, it's extraordinary to think back to the first decade of this century when more than 90 per cent of local authorities had their own local Agenda 21 strategy, with costed action plans, based around 100 integrated sustainability targets, and their own team of sustainability experts. All of this was monitored by the Audit Commission to help combat the kind of greenwashing we've become so familiar with since then.

The favourite clichés of those decades included ringing exhortations to do this or that 'in the name of our children and our grandchildren'. These nameless, faceless representatives of the future were wheeled out with such cornball regularity that the Sustainable Development Commission found itself reminding Ministers that doing things more

'I don't know how things are going to turn out, whether we get to live in the kind of healed world where there is a more human-oriented system that prioritises care and happiness and health above profit. But you have to think about the kind of sacrifice that this might mean. Right now, we're understandably upset about people going to jail for a few years – those four- and five-year sentences have really shocked people. But that's just like the beginning – a little step on the road, especially when you think historically how many bodies are already piled up on that road to have got us to where we are today. It's not easy thinking about the sacrifices we're going to have to make in the future.'

Daniel K.

'I wouldn't have it any other way, though I am sorry about the impact it's had on my parents. It's not been easy for them, although they are supportive now. As it happens, I would attribute to them my strong feelings for justice and equality, and to the way they brought me up. They instilled a sense of right and wrong in me from the start, and the fact that my life isn't worth any more than anyone else's, which is a big driver for me thinking about how best to respond to this crisis. But I know I'm definitely not an easy child in that respect!'

Ella

'It's difficult to describe, but I see all this as a kind of "fundamental solidarity", bridging so many cultures and generations. And that makes me very emotional, knowing that so many people have already sacrificed so much historically, standing up for the same basic freedoms as we are today, standing up for everything we know and love. At the same time, what makes it all the more intense is knowing that everything we know and love is now collapsing. Those two things coming together are very powerful.'

Paul

10: INTERGENERATIONAL JUSTICE

sustainably was just as important for the present generation as for future generations!

However riddled with jargon, or even insincere, that rhetoric may have been, it did at least remind decision-makers, in business and civil society as well as in government, that they should be much more mindful of the interests of generations to come rather than ruthlessly prioritising the present at the expense of the future. The explicit advocacy of 'justice between generations' that lies at the heart of genuinely sustainable development is so much more proactive than the default assumption that the interests of future generations are best served simply by improving the material standard of living of the present generation. And the default assumption underpinning that default assumption is that all we need to ensure that outcome are higher levels of economic growth, year on year, indefinitely into the future.

Those assumptions still run very deep – not least because they seem to have served all industrialised nations pretty well since the Second World War. Each new generation could indeed count on improved living conditions, a higher standard of living and more advanced technology. The fact that governments, of both parties, were essentially 'cooking the books' throughout that time was overlooked by all and sundry. Nobody seemed to care very much that the prosperity being generated was only possible by externalising the costs of that growth onto the environment, the climate and, of course, onto future generations. Even now, it's still only a minority of economists who accept that today's continuing obsession with achieving economic growth is causing greater costs for society than it is generating benefits.

These costs, already being experienced today, let alone in the future, impact all generations, but not necessarily in the same way, and not just from an environmental point of view. Impacts on human health reveal another startling example of the way such externalities give rise to chronic intergenerational injustice.

In November 2024, the UN's Food and Agriculture Organisation (FAO) released its latest 'State of Food and Agriculture' report, assessing

'I know that my journey to this point has been a bit different – I didn't start off in Extinction Rebellion and then "graduate" to Just Stop Oil! For me, it's always been about some kind of class struggle, seeing sustainability as the best way to help poor people, particularly those disproportionately affected by the climate crisis. I love nature now, but I'm a city girl at heart, born and raised in London, studying robotics at London Met, with a single mother who moved here from Morocco.

'So we talk a lot about what's going on in North Africa and Southern Europe – right now, my grandma in Morocco is affected every day because of drought conditions, with just a drizzle of water coming out of their taps. And all the bathhouses shut down!

'My mum really gets it, but, as is the case for a lot of other activists, just keeps asking why it has to be me! "Why are you the one who has to be arrested?" If I ended up going to prison, she'd still support me, but she'd be really annoyed! And this is a two-way street – I like to think that I'm radicalising her a bit. She's the one who keeps me up to date with what's happening in Morocco and elsewhere because of the climate crisis.'

Hanan

'We both feel that our responsibility as parents has to come first. With two children under the age of ten, it just doesn't seem right to be following a course of action that would end up in long prison sentences. But this is difficult. In many ways, I know that I should be using my "white privilege", relative to other people in society today, to do more, especially as I'm comfortable, personally, with the idea of going to prison. And I feel called to do this work precisely because of my children. At the same time, I really don't want to end up in prison now – precisely because of my children!'

Daniel H.

10: INTERGENERATIONAL JUSTICE

both the direct and indirect costs of today's global food system. It led with this utterly astonishing statement: 'Agri-food systems generate benefits for society, but also have over $10 trillion in hidden costs for environmental, social and economic sustainability globally.'

The biggest component of that $10 trillion is accounted for by 'unhealthy dietary patterns', i.e., what we're all eating. People's diets are increasingly shaped by their dependence on the kind of ultra-processed foods that make huge profits for Big Food. At the same time, a report from the Food, Farming and Countryside Commission here in the UK showed that the UK's addiction to ultra-processed foods, favouring fatty, salty and sugary ingredients, is costing the country £268 billion a year in direct and indirect health impacts. It's important to remember, by way of contrast, that total spending on the National Health Service in the UK amounts to about £220 billion a year.

All generations are being harmed by the failure of governments to regulate today's out-of-control global food companies. But young people are getting a particularly raw deal. There is now growing evidence that dependence on ultra-processed food is addictive, with young people's taste-buds effectively 'hacked', reinforced by seductive advertising and marketing campaigns, 'normalising' cravings for chemically-enhanced 'empty calories', devoid of any nutritional value. This heightens their risk of future health problems, including diabetes, obesity, coronary heart disease and so on, while governments and regulators look on impotently from the sidelines, reluctant to do anything to threaten the economic growth generated by Big Food.

This is just one of many areas where young people are beginning to realise that their quality of life will almost certainly not be better than that of their parents. High rents and unattainable mortgages, student debt, the cost of living, air pollution, poor water quality, mental health issues – all were raised in my conversations for this book, even before we started talking about the Climate Emergency!

OLLIE SWORDER

By the time *Love, Anger & Betrayal* is published, I hope to have finished writing a story of my own. There are a lot of people involved in Just Stop Oil with a very similar story to mine, increasingly frustrated that our friends and family just don't see things the way we do! For me, that's particularly the case in terms of my relationship with my mother.

So I wanted to find a way of telling a story through two lenses: both from the perspective of an activist, *and* from the perspective of their mother. I want to provide my own authentic, empathetic lens on the lives of climate activists, rather than the story so often told by the news media.

The easiest bit to sort out has been the fictionalised accounts of my actual experiences: arrest, troubles with mental health, experience with love, conversations with other activists, internal quandaries about taking action ... I found this all flowed straight from my soul when I described the discomfort of a police cell and the viscerally uncomfortable experiences of arrest – including the pain of handcuffs! – when I am, in all other circumstances, an entirely obedient person!

Creating a genuinely authentic 'activist' character is relatively simple. But the other half of the challenge has been to represent the parents of activists equally empathetically and authentically. I'm continually trying to understand all the feelings that might have caused my mother's negative reaction and subsequent changes in our relationship. Feelings of loss? Thinking that I've thrown away my future? Scepticism about my new activist friends? Feelings of betrayal? Fear of what I'll do next? It sounds crazy, but I'm hoping to write this in such a way as to take the mother's character on a journey that I wish my own parents could go on, coming to a place not only of understanding and support, but active engagement with protesting.

Finally, the actual storyline that links these dual perspectives is a shared experience of the loss of a loved one due to a record-breaking heatwave here in the UK at some point in the next couple of years. I genuinely worry about my grandparents being very vulnerable in this way. I still don't experience a lot of people understanding the severity of the dire situation we're in.

I feel I've always been 'a scientist'. Then, because of the Climate Emergency, I ended up as an activist. And now I'm hoping to become a storyteller.

10: INTERGENERATIONAL JUSTICE

NOT SUCH GOOD ANCESTORS

All this, and more, is happening on our watch, as notionally 'responsible adults'. All this goes largely unremarked, with our 'moral compass' set to one side, conveniently out of sight. Any thought that we collectively, as parents, grandparents and older people, might be justified in thinking of ourselves as 'good ancestors' becomes more and more preposterous with each passing year.

Roman Krznaric chose *The Good Ancestor* as the title for his 2020 book – still one of my all-time favourites – based on his admiration for a man called Jonas Salk, a brilliant medical researcher:

> In 1955, after nearly a decade of painstaking experiments, Salk and his team developed the first successful and safe polio vaccine. It was an extraordinary breakthrough; at the time, polio paralysed or killed over half a million people worldwide each year. But he was not interested in fame and fortune – he never sought to have the vaccine patented. His ambition was 'to be of some help to humankind', and to leave a positive legacy for future generations. There is no doubt he succeeded. In later years, Salk expressed his philosophy of life in a single question: 'Are we being good ancestors?'

And the principal reason why we are not being very good ancestors at the moment, in Krznaric's view, is that we've allowed our 'marshmallow brain' – fixated on short-term desires, impulses and rewards – to dominate our 'acorn brain', which helps keep the future in mind as we work towards long-term goals. As a result, all generations, across the whole of society, are subject to 'the tyranny of now'.

The Good Ancestor proposes a host of ways to combat the drivers of short-termism and to develop both the mindset and the practice of 'thinking long'. Powerfully reminding us, on the climate front, that 'we are the first generation to know that we face unprecedented global environmental risks, but at the same time we are the last generation with a significant chance of doing something about it', in the words of Johan Rockström, one of the world's leading climate and earth-system scientists. He sets out to demonstrate why there is still so much we can do to avoid our otherwise inexorable descent into climate breakdown.

After nine years as Chair of the Sustainable Development Commission, providing independent advice to the Labour Government between 2000 and 2009, I suspect I may feel rather more jaundiced than Roman Krznaric about the readiness of politicians to address that challenge of chronic short-termism. But even my scepticism on that score is as nothing compared to today's Just Stop Oil activists! Without exception, they were dismissive of the record of the previous Tory Government in balancing the interests of different generations, and only marginally less dismissive of today's Labour Government. They see its balancing act on new fossil fuels – committed, on the one hand, to 'no new oil and gas assets in the North Sea' and, on the other, refusing to withdraw approval for 100 such licences granted by the Tories, including the vast Rosebank development – as despicable hypocrisy. Whilst acknowledging that the 2008 Climate Change Act, the first such legislation brought in anywhere in the world, has had some beneficial impacts, and that the UK has no worse a climate record than any other leading industrial nation, it's still the almost complete lack of urgency and proportionality that has driven them to civil disobedience.

There's certainly nothing on the Statute Book for the UK as a whole that explicitly recognises the importance of Intergenerational Justice. The citizens of Wales, by contrast, are now celebrating the 10th Anniversary of the 2015 Well-being of Future Generations Act, which requires all public bodies in Wales, not just the Welsh Government, 'to think about the long-term impact of their decisions, to work better with people, communities and each other, and to prevent persistent problems such as poverty, health inequalities and climate change'. And they have their very own Future Generations Commissioner for Wales to ensure that the Act is being fully and accountably implemented.

10: INTERGENERATIONAL JUSTICE

USING THE LAW

However, having laws in place is one thing; ensuring they're properly enforced is altogether another. In the field of climate change, this has led to an extraordinary proliferation of judicial reviews and legal actions of one kind or another, holding Ministers and governments to account for their perceived failures to exercise their powers responsibly or to abide by the laws of the land. Looking back over the past fifteen years, this is probably the biggest single addition to the tactical armoury deployed by climate campaigners.

According to Columbia University's Sabin Centre for Climate Change Law, there are now more than 2,500 climate lawsuits around the world. Using the law is neither quick nor cheap, but it can be very effective. A report from the London School of Economics (LSE) analysing more than 500 recent lawsuits outside of the USA found that 55 per cent of cases had what it describes as a 'climate-positive outcome' – and though this is much harder to assess, the report also indicated that many of these cases had had a clear impact on public policy.

Given the huge costs involved, NGOs generally only pick up the legal cudgel as a last resort, usually to try to stop bad things from happening, or to force the Government, or its agencies, to carry out its obligations in law. After ten years of relentless legal process across four presidential administrations, 'Juliana vs United States', perhaps the best known lawsuit brought by young people against the administrations that continued to betray them, ground to a definitive halt when the US Supreme Court finally closed it down. It's an extraordinary story (with the whole saga covered in detail on the website of Our Children's Trust: https://www.ourchildrenstrust.org/), which inspired more than sixty youth-led climate lawsuits worldwide, including some substantive victories.

The outcome itself does indeed matter, but even when a particular lawsuit fails, it can create a really powerful narrative which then persuades decision-makers to intervene more effectively. For instance, the report from the LSE includes the example of the Australian teenager, Anjali Sharma.

'I recognise that I've been able to take this path coming from a very privileged place, being white, middle-class, having a huge amount of love and support around me, and not having any dependants. And I know that's not the case for huge numbers of people who have to prioritise other goals and other people.

'But I still think a lot of other people are making excuses and really aren't asking themselves the hard questions about what it is that's holding them back. I know it's important to go on being compassionate about this, knowing how scary it can be to jump in, but I know people who have sacrificed their careers, their relationships and the houses they live in. Precisely because of their concern for the future.'

Phoebe

'I'm thinking a lot about what happens after I'm released. Up until now, the narrative of the climate crisis has been at the centre of what I've been doing, and I have no illusions that ecological collapse is the most existential threat to all life on the planet and will continue to make all social injustice that much more pronounced.

'At the same time, I've come to realise that it's the work of justice in all its breadth that is at the root of it all, and that in the long term, this is what I am committing my life to.

'I used to think that staying in education might provide me with some stability from which to continue doing this work – and that is the case for some people I know. But I've realised now that it is not the environment of institutional education that I need and that, for me, stability will come from a more spiritual source.'

Cressie

Her attempt to force the Australian Government to acknowledge that it had a legal 'duty of care' to protect young people from the future ravages of the Climate Emergency did not succeed. But it persuaded David Pocock, an independent Senator, to introduce a Bill requiring federal and state governments to consider the well-being of current and future children when making decisions that contribute to climate change – very much like the Well-being of Future Generations Act in Wales.

In Europe, six young Portuguese activists brought a lawsuit against the thirty-two governments of the EU which was heard at the European Court of Human Rights (ECHR) at the end of 2023. It failed, but as part of the same hearing, the ECHR top bench ruled that the Swiss Government's inadequate emissions-reduction strategy violated the rights of a group of older Swiss women to family life and to better protection against climate-induced impacts. This landmark verdict exposes all forty-six members of the Council of Europe to similar cases being brought in their national courts.

The huge uplift in awareness that these two parallel cases engendered was extraordinary, as was the case in the state of Montana in 2024, where sixteen young people successfully argued that the state had violated their constitutional right to 'a clean and healthful environment'. This probably won't have much impact outside of Montana, but it's provided a huge boost to similar youth-led lawsuits, many of which are funded by Our Children's Trust, an impressive legal NGO which has Intergenerational Justice at its heart.

RISING UP

I didn't ask Just Stop Oil campaigners about these legal initiatives. Their experience of the law is very different, as laid out in Chapter 8, and I suspect they would see in all these lawsuits the same pattern of delay, special pleading and the dead weight of a political system captured and corrupted by the fossil-fuel incumbency.

For them, the message is clear: young people have to confront that system, not comply with it. I've always felt that an inflection point of this kind, in one form or another, is inevitable. Back in 2013, I wrote a book called *The World We Made*, looking back from 2050 to show how we got to be living in a reasonably secure, compassionate and fair world by then. Albeit still massively disrupted by climate shocks of one kind or another, but 'on the mend'. One of the key – imagined! – turning points that set us on that path was this:

> On 14 July 2018, in a way that politicians had assumed would never happen, young people suddenly realised the full implications of accelerating climate change. The Enough! Movement exploded into life in France, the USA, India and Russia, and then went global. Huge numbers of young people were mobilised around a shared vision expressed in the Enough! 'Manifesto for Tomorrow', with young people occupying government buildings, parliaments, stock exchanges, newspaper and TV companies, banks, oil and mining companies, town halls and civic centres – an irresistible tide of shared fury and compassion.
>
> In the first couple of weeks, thousands were killed or injured in the riots that ensued. But that just brought more young people onto the streets. Chaos reigned for several months – at which point governments all around the world began to respond to our demands, pledging to introduce radical reforms, particularly on fair taxation, climate change and protection of the natural world.

Well, that hasn't happened – yet! But twelve years on from that fictional moment of hope, I find myself with two very tentative ways of keeping despair at bay. First, I still hanker after some kind of 'Enough!' moment in the future. I cannot imagine a world in which our brazen contempt for young people goes indefinitely 'unpunished'. I cannot imagine a world without some uprising, in which the police, state troopers or army personnel find themselves forced to choose between gunning down innocent young people or obeying the orders of their increasingly cruel and oppressive rulers.

When talking about the Civil Rights protest in Birmingham, Alabama, in 1963, interviewees for this book recalled most powerfully what became known as the 'Children's Crusade', when wave after wave of young people flooded Birmingham's streets, singing and joining arms. With its jails already filled, Birmingham's racist and domineering sheriff, Bull Connor, 'opened up on demonstrators with batons, dogs and water hoses. The scenes that resulted were horrifying. Streams from high-pressure fire hoses ripped the clothing from the backs of protesters ... doctors treated other demonstrators for dog bites', as described in Mark and Paul Engler's *This Is An Uprising*. The authorities in Birmingham backed down; within a year, the Civil Rights Act was passed.

Any intergenerational clash of this kind may not end well. The most extreme and shocking interpretation of Intergenerational Justice I've ever come across is in a novel called *The Uninvited* by Liz Jensen, founder of the wonderful Rebels' Library. Set in the near future, it invites readers to imagine what might happen if justice for young people remains as unobtainable as it is today, primarily because of the indifference and selfishness of their parents and grandparents. 'Uninvited', children start to 'take possession' of adults in key positions in industry and society, inducing them into increasingly devastating acts of sabotage. Economies grind to a halt in a remarkably short period of time – even as the children themselves regress into 'wild creatures' better equipped to survive as our industrial civilisation implodes. 'Dark' doesn't do *The Uninvited* justice! But there's a sliver of potential redemption even in this extreme scenario: nature will heal, and will be fully regenerated as the successors of the savage, 'innocent survivors' get to having another crack at building a just, compassionate human society living in harmony with that natural world.

Is that really what it will take to 'break the tyranny of now'? Paradoxically, the other way that I have of keeping despair at bay is to imagine that it's 'the tyranny of now' itself which will force this generation of politicians to change their ways. Not out of any greater concern for young people and their future, but out of concern for themselves and their own political survival.

This improbable outcome depends on buying into as dramatic a 'socio-economic tipping point scenario' as anything that is happening today in the world's oceans or rainforests or polar regions. Like all good scenarios, it starts with some hard-edged facts: the global insurance industry, which underpins the whole global economy, as nothing happens without insurance, is already showing signs of coming apart at the seams because of climate change. The true extent of the economic damage being done by climate-induced disasters of one kind or another already reveals a 'non-viable business model' or 'an uninsurable world' – to use two recent quotes from insurance industry leaders. So, just imagine, as happens for real with all physical tipping points in our climate systems, such as melting sea ice and rising ocean temperatures, if one thing led inexorably to another.

It could start anywhere, but let's take Florida, as they've probably got it coming to them anyway! Imagine the 'worst-ever hurricane season', in 2025 or 2026, causing in excess of $1 trillion of economic damage. The figure of $1 trillion may sound completely outlandish, but as I mentioned on page 113, super-storms and extreme wildfires, including those in Los Angeles in January 2025, are already causing hundreds of billions of dollars of damage.

What would happen then? Imagine two thirds of state-based insurance companies in Florida are immediately bankrupted; the big nationwide insurance companies start to withdraw completely, 'red-lining' the entire state, which then has to raise tens of billions of dollars every year as the 'insurer of last resort'. But it turns out to be too late for such drastic interventions. 'Contagion' in the market drives investors to exit the whole insurance industry, with company after company going to the wall; the reinsurance industry, which insures the insurance companies themselves, comes under massive pressure, as surely as melting

10: INTERGENERATIONAL JUSTICE

ice follows warming oceans, and has to be bailed out by the World Bank, the ultimate insurer of last resort. The rest, as they will surely say at some point in the future, is history. The economic disruption and suffering would be horrendous. Tens of millions of people would find their lives brutally upended, even more painfully in the rich world than in poorer, but more resilient, countries.

Crazy? Of course. But that's the only way, as I see it right now, in which today's suicidal capitalist system turns out, against all the odds, and at absolutely the last possible moment, to be capable of rescuing itself from itself.

Let's just hope young people rise up against that system long before this kind of scenario comes to pass.

'I was lucky being brought up believing that activism and protest is a good thing, and something we should all be involved in as a fundamental part of our democracy. I was always encouraged to stand up for what I believed in, to use my voice, one of the most powerful tools we all have. When I was little, I was just obsessed with the Suffragettes – I was the biggest, gobbiest, most angry wee feminist ever!

'My parents were very clear that you shouldn't waste things, but you should appreciate everything you've got, not buy new things unnecessarily. My upbringing centred around not wasting/being grateful/looking after what you've got/just having what you need. And celebrating the planet – celebrating its beauty and its treasures.'

Niamh

'It still seems perfectly possible to me that young people will bring down the system, as happened in the Arab Spring and with other revolutions in the past. And it will, of course, be complete chaos along the way – unless we've prepared people in advance for the inevitability of doing things very differently'.

Sam

PHOEBE PLUMMER

After something of a messy childhood – comfortable, middle-class, Tory-voting family and a climate-denying dad, divorced parents, illness etc. – I was 'adopted' by my then partner's supportive and loving mother into a family with strong social justice values and a history of radical campaigning – a world away from my biological family! I found myself there, discovered what I really cared about, tried to make things work at uni, but then got completely caught up in climate activism.

INVOLVEMENT

I was fascinated by both Extinction Rebellion and Insulate Britain, but always seemed to have some reason not to commit to taking action. That eventually happened with Just Stop Oil in August 2022, with three arrests after a week of action disabling petrol pumps, and I've been arrested several times since then, for slow marching; for throwing soup, together with Anna Holland, at Van Gogh's Sunflowers and other actions. It's full time, whether I've been in prison, or free to confront those in power threatening everyone's freedom in the future.

MOTIVATION

I want to be living a life that is completely aligned with my moral values – and I really don't take it for granted that I'm able to live the way I do today only because of all those who took action to protect our rights so many times in the past. Climate breakdown now threatens all that, and a whole lot more.

INSPIRATION

We live in a society where we are made to feel powerless to bring about systemic change against injustice. Anyone who has overcome that and has come together in collective action to fight for a better world, is a true inspiration to me, no matter what kind of action they've taken. Each and every person who has been placed in handcuffs next to me, and those who've waited outside police stations all night to greet me on release, each person who has faced trials with me, and those who have hugged me after a guilty verdict is returned, every person who has also given up their liberty and been to prison, and those who cared for me when I've left prison, not only inspire me but make it possible for me to take the actions I have. It would fill a whole book to name all these brave, beautiful people!

IN NATURE

Being near or, even better, immersed in any natural body of water fills me with a feeling of foundational peace.

QUOTATION

'There is a time when the operation of the machine becomes so odious, makes you so sick at heart, that you can't take part. You can't even passively take part. And you've got to put your bodies upon the gears and upon the wheels ... upon the levers, upon all the apparatus, and you've got to make it stop And you've got to indicate to the people who run it, to the people who own it, that unless you're free, the machine will be prevented from working at all.'
Mario Savio, of the Berkeley Free Speech Movement

RESOURCES

I've picked two books: The Ministry for the Future, by Kim Stanley Robinson, because I've never read a novel with such realistic depictions of the grief and suffering caused by the climate crisis, yet balanced by realistic hopes of effective global action.

And This Is an Uprising, by Paul and Mark Engler, as an accessible, comprehensive guide to why non-violent civil resistance is the way ordinary people can make this global action happen.

WHAT LIES AHEAD?

I used to think that taking part in direct action would end in a huge moment of victory when hundreds of us would leave police cells to the news that we'd overwhelmed the system, and the Government had announced some landmark climate policy as a result. It turns out that's a fairy tale!

So, I think the next few years will be filled with many wins and many losses as we keep pushing for the necessary climate action and incrementally moving in the right direction. I think we'll see more public awareness about the severity of the climate crisis, and acceptance of the urgent need for change, combined with the usual lies peddled by our corrupt media. I think we'll see more people stepping up to resist genocidal government policies, despite draconian repression of protest increasing. And, always, we'll be building strong, resilient communities to face whatever happens together.

ROSA HICKS

I've been passionate about social justice since I was a teenager, and am very involved in campaigning around refugee rights and anti-austerity. I then realised, when I was at Manchester University, studying psychology, that the climate crisis is so grave that however much progress we make on these things, it will all be swept away as the breakdown worsens. I saw that for myself living in Australia for four years, which led to me returning to the UK in 2023, to work with Just Stop Oil.

INVOLVEMENT
2018–19: limited engagement with XR in Manchester.
2019–23: supported various campaigns in Australia.
May 2023: Chelsea Flower Show action.
March 2024: arrested for conspiracy to commit burglary – no charges brought.
July 2024: arrested for conspiracy to interfere with critical infrastructure – Heathrow.

MOTIVATION
The climate crisis is just so imminent. It's not like I can morally put it on the back burner telling myself, 'I'll deal with that in a couple of years.' Everything is against the clock: the longer we leave it, the worse it's going to get; the more tipping points we'll hit. For me, it is that black and white, something that has to be done, even if we're not successful. It sometimes seems almost selfish: knowing that I'm acting in line with my morals, doing what I can, regardless.

INSPIRATION
What inspires me most are the people that I've had a personal connection with. Ordinary people like myself, doing amazing, committed, self-sacrificial things. Two of my friends in Australia, Andrew George and Violet Coco, who played a big part in advising me to come back to England to get involved here, are just amazing – brave, determined and willing to sacrifice anything that they need to. They've both been to prison and were some of the first people in Australia to be locked up for climate protest. Back here in England, it's my friend Phoebe Plummer – seeing that complete resilience of hers, having a friend by my side who's just so powerful yet humble.

IN NATURE
The place in nature where I feel secure and at peace is just with water –

whether it's the sea or a lake or a river. It brings me calm, happiness, mindfulness. In Australia, I used to have the sea at the end of the road. Whenever it was all a bit much, I could just go down and wade into the water, and it brought everything back to a steady level.

QUOTATION
I thought a lot about this, and it has to be Margaret Mead's: 'Never doubt that a small group of thoughtful, committed citizens can change the world; indeed, it's the only thing that ever has.'

That's so important to me, because a lot of the time it feels like we're just so small and so isolated.

RESOURCES
I would one hundred per cent recommend The Ministry for the Future by Kim Stanley Robinson. A lot of it's quite depressing – especially the start! – but it also shows that we might actually get our act together eventually. Even at the last minute!

WHAT LIES AHEAD?
I do hope in a realistic sense that there will be a form of real change – more than the government just declaring a Climate Emergency like they did six years ago – to get the cogs turning. Much like in the Second World War when suddenly every single part of the economy was converted into addressing that crisis. It's possible. It can – underline can – happen!

(See p. 266 for the statement from Rosa and her co-defendants after their trial.)

(In May 2025, Rosa was given a fifteen-month custodial sentence, deemed to have already been served by time spent on remand, and required to contribute £2,000 to court costs. Phoebe was given a twenty-four-month custodial sentence, suspended for twenty-four months, and required to contribute £500 to costs. She still faces a criminal damage re-trial this year, and a public nuisance trial in November 2026.)

11: WHAT LIES AHEAD?

'If people insist on living as if there's no tomorrow, then there really won't be one.'
Kurt Vonnegut

I THOUGHT I should try to answer the same question that I've asked of the young activists I interviewed in co-creating this book. Like most of them, I'm really not sure what lies ahead.

Some of my co-authors cared more about the fate of Just Stop Oil than others. Loyalty was strong with almost all of them, especially for those who were not previously involved in Extinction Rebellion or Insulate Britain, but all recognised that Just Stop Oil was no more – and no less – than a temporary vehicle, driven by a powerful core of ideas, a steadfast commitment to the science of climate change, an ethos, the resolute practice of non-violence and a passion for justice.

Throughout the past year, I've kept coming back to Martin Luther King's well-known reference to the 'arc of the moral universe' and his belief that, over time, it 'bends towards justice'. That belief has been sorely tested over the past year! Indeed, it's been a massively problematic year in which to write a book like this. Donald Trump is back in the White House, ruthlessly hacking back critical support for tens of millions of US citizens; summarily undoing decades of steady progress on human rights, LGBTQ and trans issues, while Diversity, Equality and Inclusion (DEI) are treated as subversive viruses; and Environmental, Social and Governance (ESG) is portrayed as an anti-capitalist plot. And when it comes to the climate crisis, so irrational and deep-seated is Trump's denialism that any references to the word 'climate' have had to be systematically removed from government websites.

The year 2024 was not quite as bad in the EU, but in 2025 we did begin to see the European Commission tracking Trump one backward step after another. In response to the war in Ukraine, and Israel's war to destroy the Palestinian people and 'cleanse' them from both Gaza and the West Bank, militarism has been turbocharged, with the manufacture and sales of arms back to the 'glory days' of the Cold War.

And all that before we even begin to talk of stalled progress on climate, biodiversity, widening inequality the world over, plastic pollution, toxic chemicals, ultra-processed food, global poverty and so on and on. All the 'bending' has been devastatingly in the wrong direction, away from a more just and sustainable world.

What that has meant for me, personally, is that it's the twenty-six young activists, for whom I've tried to provide a platform in this book, who have kept hope alive. I don't want to overdo this: I'm never going to give up on staying hopeful, somehow or other. If I'm still alive in 2040, at the grand old age of ninety, and the average temperature increase has crashed through the 2°C barrier, which is now all but inevitable, and the lives of hundreds of millions of people are being hammered by ever more frequent, ever more intense and ever more devastating climate-induced disasters of one kind or another – again, all but inevitable – I'll still be urging people to double down on whatever needs to be done to prevent it from getting even worse. As Nick Cave puts it, 'Hopefulness is not a neutral position either. It is adversarial. It is the warrior emotion that can lay waste to cynicism.' Working with this particular group of young 'warriors' has been extraordinarily re-energising. No cynicism here!

By 2040, they'll be in their late thirties or forties, with – or, I have to suspect, mostly without – children of their own. And I will be as scared for all of them as I am for our own daughters.

By virtue of the actions they've taken through Just Stop Oil, and their certainty that only continuing civil disobedience will force today's politicians to grasp the reality of accelerating climate change, they've all had robust – and often very painful – conversations with their parents and siblings. And they've discovered, from similar conversations they've had with friends, that this just doesn't happen in most families – which made me wonder how many people reading this book will have found a way of talking about all of this with their children / grandchildren / parents / grandparents / siblings / partners / best friends / work colleagues? Far fewer, I fear, than the crisis demands. If this book achieves nothing else, I hope it will have opened up more of those conversations. As well as providing a more balanced view of what Just Stop Oil was, and particularly its young activists are, trying to achieve.

Apart from Martin Luther King, my other go-to historical companion over the past year has been Emmeline Pankhurst, a figure who features prominently in the kind of historical analogies that all direct action campaigners refer to. As I mentioned in Chapter 7, I was impressed at the sophistication with which interviewees handled the nuances between, on the one hand, their deep admiration for everything the Suffragettes went through while, on the other, emphatically distancing themselves from the Suffragettes' resort to violence.

The Women's Social and Political Union relied heavily on an appeal to future generations; its supporters argued that even if its critics at the time failed to understand 'the necessary evil of strategic violence', future generations would, in retrospect, as they enjoyed all the benefits of universal suffrage, praise the difficult choices the Union had had to make, including, if necessary, 'the laying down of our lives'. Emmeline Pankhurst put it like this in her autobiography, *Suffragette*:

> We are fighting for a time when every little girl born into the world will have an equal chance with her brothers, when we shall put an end to foul outrages upon our sex, when our streets should be safe for the girlhood of our race, when every man shall look upon every other woman as his own sisters. When we have done that, we can rest upon our laurels, assured that we have passed on to future generations an inheritance worthy of the great human race of which we are humble members.

'I still don't know why I ended up on the path that brought me here, when so many others in similar situations didn't. It would be stupid to claim that it's because of any kind of "moral superiority" – it's absolutely not like that. In fact, I feel lucky to have met people who encouraged me to get involved. And I've also learned that you have to be prepared to put up with a lot of sadness and grief.'

Olive

'I feel very strongly that prison is where I need to be – indeed, the only place I could feel at peace with myself at the moment. And I know that being in prison is often harder for the people we leave on the outside than it is for us.

'And there's something here that people don't see – they focus on the actions but miss the degree to which the community of Just Stop Oil makes all that possible. I'm able to be here in prison because I have faith that I will be held by this beautiful community we've created, and that when I come out of prison, with no money, limited job prospects and a prison record – I'm not scared about that, because I know I will be held through these times.'

Phoebe

'I haven't got a fresh palette that I can just paint some kind of rosy future onto; the palette I'm looking at is already pretty ripped up, with me spending more time thinking how we can deal with societal collapse than future career prospects. To be honest, that sucks, but we're the ones who are going to have to deal with things when it all goes wrong, take care of the kids that are being born now, live through all the shittiness that is inevitably coming our way, and hopefully stabilise things at some point in the future.'

Hanan

11: WHAT LIES AHEAD?

E. H. Taylor, a strong supporter of the Suffragettes was fulsome in his recognition of their courage: 'Never forget what the verdict of posterity will be on all this. The world does not know it yet, but it will then, and then your names will be on the lips of its children's children through all the ages.'

There is something very poignant about the comparison that can be made between the Suffragettes more than a hundred years ago and the 'Radical Flank' of today's climate movement – while pointing out that 'ensuring their names are on the lips of future generations' is literally the last thing on the minds of the activists I've been spending time with! But there's a deeper significance to what they, and everyone involved in Just Stop Oil over the past three years, set out to achieve, and I hate the way this has been almost completely ignored.

I know that some will regard this as being almost self-indulgently naïve, but the young people whose views you've been connecting with, chapter after chapter, really do believe that if people could just 'see' what they see, could just 'step into the shoes' of people in the second half of this century in the way they do, and could be helped to understand that we still have time to avoid the worst consequences of climate breakdown, then they'd be out there forcing our elected representatives to get a grip on the Climate Emergency in the same way as they did, more or less, on the Covid pandemic.

Not by everyone gluing themselves to motorway slip roads, or disrupting sporting events, or hurling soup at the screens protecting treasured artistic icons, but by exercising their rights not just to vote every now and then, but to speak out, to protest, to shame all those blocking this necessary revolution, and to organise, street by street, house by house.

Naïve? Of course. But how could it be anything else in a world where authentic, genuinely hopeful idealism has been entirely driven out of politics?

'Right from the start, what I glimpsed in Just Stop Oil was a different way of living, the way people talked with each other, connected with each other – it felt like a real community. Almost immediately, I felt a real sense of belonging. I think this is the most important thing about resistance – that sense of community, belonging and support. And I guess it makes a lot of sense that this can blossom in spaces of non-violent resistance given the work we're doing. We certainly can't rely on state structures to look out for us, because this work is all about resisting the state.'

Cressie

'It all comes down to what people see as a responsible or sane responses to what is happening in the world. For me, what we're doing with Just Stop Oil is the most logical, sane thing to be doing. What is completely *insane* is pretending that we can keep on living normally, with our crazy, consumption-driven lifestyles, even as the climate scientists keep on telling us that we're crashing through one threshold after another.'

Ella

'It's so easy, with people who aren't involved, to feel like some kind of weird doomsday preacher, especially if you know they don't really want to hear about the climate crisis. You become the odd one out. But when you're with JSO friends and fellow protesters, it just feels much more normal to have these worries and to be prepared and able to face up to them.'

Oliver

11: WHAT LIES AHEAD?

FREEDOM RIDERS TO CLIMATE CAMPAIGNERS

This is why it isn't actually the Suffragettes who come to mind when I think about Just Stop Oil's young activists; it's the Freedom Riders who played such a critical part in the Civil Rights Movement in the USA in the 1960s.

It would be entirely inappropriate to make any direct comparison here. Whatever sacrifices Just Stop Oil activists have made, however draconian the law of the land may have become in suppressing their fundamental right to protest, and however hateful the right-wing media may have been in their vilification of them, that is as nothing compared with what the Freedom Riders went through. Just Stop Oil and Youth Demand activists are not on the receiving end of any organisation as hateful as the Ku Klux Klan. They are not routinely hospitalised because of the violence done to them. They did not feel obliged to write their wills before committing to the next action.

As I write this, on their inspirational Fighting Oligarchy Tour, Bernie Sanders and Alexandria Ocasio-Cortez are powerfully reminding US citizens of this heritage, summoning up the memory of Congressman John Lewis, from Georgia, who died in 2020. He was one of the original Freedom Riders, assaulted on countless occasions, and savagely beaten on the Edmund Pettus Bridge in Selma, Alabama. His rallying cry in those days was 'Get in good trouble, necessary trouble, and help redeem the soul of America.' Bernie Sanders and AOC are now breathing new life into that powerful exhortation.

'Get in good trouble, necessary trouble.' I don't think Just Stop Oil ever relied on any one slogan, but that would have done as well as any.

Not least as there are so many moving resonances between its young activists and those Freedom Riders, who were also mostly young. For one thing, as touched on above, I've thought a lot about those conversations with parents, then and now, as they come to terms with the emotional maelstrom of their children putting themselves in such jeopardy – the self-questioning, the fear, the incomprehension, the anger, the pride.

'For a long time, I lost my sense of self. I'd been living on autopilot since the soup action because my identity had been stolen by the mainstream media, by the dehumanising intensity of social-media responses. All of a sudden, I wasn't Anna – I was Soup Thrower, and I didn't know how to be that person. Who was this person with death threats, fan mail and two interviews a day with international outlets? I didn't know them, and I didn't want to be them. So, as a sort of defence, I stopped being anyone. Interviews became a robotic repetition of the lines I knew would have the biggest emotional impact: lilting my voice at the right moment; letting my eyes fill with tears, or my voice rise with anger, when appropriate. I didn't recognise the person I was on those interviews, and I'm sorry I didn't take care of them better.

'Now, I'm finding myself again, and I love who I'm finding. Through my writing, my friends, my learning, I'm becoming whole again, and leaving behind that auto-piloted person of the past two years.'

Anna

'It's hard to pin down, but I definitely feel like I've had to give some things up for this life. Early on, it just made most of the relationships in my life very difficult. I was quite young then, just nineteen, and that was hard. I had to take on a lot of responsibility very early on, first with Just Stop Oil, and then as co-founder of This is Rigged. I didn't really want that responsibility; I felt that I was missing out on some of the things that I would otherwise have been doing – the more normal part of life.'

Eilidh

11: WHAT LIES AHEAD?

And then there's the bemusement that I explored in Chapter 2 – with the vast majority of their peer group, let alone society as a whole, unable to see things the way they do, despite such a clear-cut moral imperative. 'We keep trying to understand the world of people who don't want to understand the world as it is. But why don't they want to understand? How can there be such an absence of empathy?' First-hand testimonies from some of the Freedom Riders speak to that same astonishment, and anger, at the lack of empathy for fellow African-American citizens back in the 1960s.

Reading those testimonies, there's also a particularly powerful comparison between the disdain they had for the huge numbers of 'white liberals' in the 1960s, enthusiastically signed up to the cause of civil rights, but always, somehow, still standing by when push came to shove, and some of the climate movement's Radical Flank expressing similar disdain for today's 'mainstream environmentalists' always, somehow, still standing by. That sounds harsh, but just think about it.

Lastly, there's a shared 'irrational hopefulness'. If the Civil Rights Movement in the US in the 1960s had failed, the spectre of mass racial violence was seen as a very real possibility. The prospective consequences of today's global campaign for climate justice failing are even more profound. But both the Freedom Riders and Just Stop Oil/Youth Demand activists hung on then and hang on now to a deep, 'against all odds' sense of hope, regardless of immediate, or indeed any, prospect of success.

Iris Dement's song 'Workin' on a World' beautifully captures this:

> Now I'm workin' on a world I may never see
> I'm joinin' forces with the warriors of love
> Who came before and will follow you and me
> I get up in the mornin' knowing I'm privileged just to be
> Workin' on a world that I may never see.

ROSA HICKS

On 20 March 2025, Rosa Hicks was found guilty of conspiracy to cause a public nuisance at Heathrow, together with seven other defendants (see p. 256). This was their statement after the verdict:

> We thank the jury for their service and accept their decision. We recognise the constraints they were under given that the judge removed all legal defences, ruled the climate emergency to be 'irrelevant', and forbade us from mentioning that a jury has a right to acquit a defendant as a matter of conscience.
>
> Some of us now face many months in prison for planning an action that never happened. We sought to get media attention so that we could explain the growing suffering and the horror of our heating world and the urgency for global action. In that, we count ourselves successful. A small victory won in the wider struggle against complacency, false hope and denial.
>
> We have no regrets. We planned our campaign with care, aiming to avoid harm and with the intention of preventing greater harm. The bigger crime would have been not to act.
>
> Civil resistance to a morally bankrupt political class is not only necessary as an act of self-defence, it is also morally justified. There are many who know the horror of our situation who nonetheless are carrying on with business as usual, in the mistaken belief that someone else will solve the problem. We are sorry to be the bearer of bad news, but if you don't stand up and do something, we are going to lose literally everything.

11: WHAT LIES AHEAD?

THE WHOLE TRUTH

There is one highly significant difference between the 1960s and the 2020s. Despite the agonisingly slow struggle that led up to the Civil Rights Act in 1964, the US economy at that time was in the middle of an extended period of Keynesian growth, with net wealth distributed on a significantly more equitable basis than today, and democracy was functioning relatively robustly. Sixty years on, democracy in many parts of the world is increasingly threatened by the rise of populist, authoritarian and far-right political parties, at the same time as the global economy is increasingly dominated by the malign influence of hundreds of billionaires and a handful of out-and-out oligarchs accountable to no one but themselves.

I keep coming back to those 2,769 billionaires, particularly the dozen or so oligarchs who control sprawling media empires, both in print and digital, in the US, UK, China and India. It's been more than thirty-five years since Edward S. Herman and Noam Chomsky published their extraordinary book *Manufacturing Consent: The Political Economy of the Mass Media*, in which they argued that 'the mass communications media of the US are effective and powerful ideological institutions that carry out a system-supportive propaganda function by reliance on market forces'. Chomsky has spent a lifetime warning of the risks to democracy when that 'nexus' between politicians, business elites and the mass media is able to operate beyond regulation and democratic accountability. Tragically, that's exactly where we've ended up today.

Without exception, today's media empires constitute a massive barrier to addressing the climate crisis in a timely and proportionate way. The science of climate change would appear to be completely irrelevant to them, operating as they do along a continuum of contemptuous scepticism to outright denialism. They openly and persistently lie about what is going on out there on the front line of climate change, disregard the implications of what this all means for the economy – the shocking report from the Institute and Faculty of Actuaries that I

'I often find myself wondering what it will actually take to bring about the change that is necessary. I know that at some point in my lifetime there will be enough social unrest that change will happen, but I don't know whether that's going to come in time, whether people will eventually be shocked into action. Perhaps by some unprecedented climate disaster – or some combination of different things.'

Ollie

'It's our values that tell us what is or isn't important about how we should lead our own lives, and what kind of change we should be seeking. All social movements need a good strategy, of course, but they also need a vision; without that sense of shared identity and shared purpose it's difficult to work effectively together. This is what people refer to as "the ethical basis of civil resistance", and I've always felt that there is a spiritual dimension to this – although I know some people don't like using that word "spirituality"!

'But there's just so much that is wrong with what we refer to as "Western civilisation", with its endless focus on growth and consumption, and little recognition of people's deeper feelings – what it means to be human, to lead simpler, genuinely fulfilled lives at this point in history. We have to get so much better about asking these questions, inviting people to think more deeply about the spiritual aspects of our lives – and that has always been absolutely critical for me.'

Eddie

referred to in Chapter 1 was almost completely ignored – and when it comes to climate protest, both on the Moderate and the Radical Flank, their default options are to downplay it all, ridicule it or, as Just Stop Oil routinely experienced, viciously demonise it. Is it any wonder then, returning to that critical question – 'Why don't people understand?' – that so few people see the issue in the clear-sighted way that climate campaigners see it today?

I wasn't all that surprised when Just Stop Oil 'hung up the high-vis' in March 2025. Few, if any, campaigning organisations have had to deal with the vitriolic hostility directed at them by the media and by mainstream politicians intent on obscuring the reality of climate breakdown at any cost. Just Stop Oil may have made its fair share of mistakes, but I still have huge respect for the organisation and greatly admire what they managed to achieve in just three years. And I remain deeply disappointed by all those mainstream climate campaigners and environmentalists who never spoke up in support of Just Stop Oil

There may well be a 'climate majority' out there, just waiting for the right moment to show how much they care, to demonstrate how determined they are to see their elected representatives get a grip on this crisis. But I've spent more than fifty years trying to reach out to that majority of citizens, if only to mobilise a bigger minority of them, and I have no illusions left – about both my failure and theirs. If we continue to rely on the same old business-as-usual theory of change, the inevitable result will be that such a majority will be mobilised only when it is already too late to make any significant difference.

I speculated, at the end of Chapter 10, that the only thing that could bring an end to this 'conspiracy of privileged and self-serving denialism' by the billionaire oligarchs is the potential collapse of the insurance industry, and the knock-on effects of this on the global economy. When, at last, the devastating physical reality of accelerating climate change can no longer be concealed, when insured costs exceed hundreds of billions of dollars every year, as so eloquently expressed by the former CEO of Allianz Investment Management in Chapter 1, the shock will be terrifying for them, and for the political and business elites that have directly and indirectly kept so many so deep in involuntary ignorance for so long. It will then be beyond denial.

It is often suggested by historians that 'civilisations fail when their elites can no longer agree on what threatens them'. Ironically, today's elites do agree on what threatens them, and that's *the truth* about the extraordinarily precarious situation that humankind finds itself in a quarter of the way through the twenty-first century. That situation could still, even now, be transformed if citizens were able to share the truth about them – the oligarchs, the billionaires, the warmongers, the destroyers of hopes and dreams; the truth about the sheer physical impossibility of relying on economic growth, stretching indefinitely into the future, to improve the condition of humankind; the truth about our total dependence on the natural world to provide both for our material prosperity and for our well-being; the truth about all the inspiring economic and political alternatives that would allow fair, dynamic and sustainable communities to thrive the world over; and the truth, most importantly, about who we humans really are, a species with a much greater capacity for compassion, empathy, mutual respect, and for co-creating a better world for ourselves with all non-human life, than any of their assiduously cultivated tropes about 'innate' selfishness, greed, competitiveness and devil-take-the-hindmost materialism would have us believe.

Ultimately, a proper understanding of those truths provides the only sure-fire defence we have against falling deeper and deeper into the very dark world that now confronts us.

This takes us back to 'The Whole Truth', to the sacrifices needed to help people see those truths for what they really are, and then to work together to co-create that better world. As has so often been the case over the past decade, UN Secretary-General António Guterres nailed this clear-cut moment of truth: 'We have a choice. Collective action or collective suicide. It is in our hands.'

We all have that choice. And it is now uncomfortably binary: action or suicide.

SAM HOLLAND

I work full time as a 'campaigner' on civil-resistance projects to force action from the UK Government on the climate and on the genocide in Gaza. I'm currently working for the Youth Demand campaign which I co-founded in March 2024. In September 2022, I dropped out of university to join resistance with Just Stop Oil, where I did youth mobilisation for eighteen months. Prior to university I co-founded a zero-waste retail business called Oat Float in Bristol, which is still trading today.

INVOLVEMENT

My first experience with climate activism was at university, where I organised with XR in summer 2022 and set up a group called Student Rebellion in Leeds. I quickly left the XR work and joined Just Stop Oil, helping to head up the youth mobilisation efforts.

I have done eight arrestable actions with JSO, from blocking oil terminals in August 2022, to climbing a gantry in November 2022, to spraying the University of Leeds with orange paint in October 2023.

I was involved in setting up Umbrella in early 2024, and out of Umbrella's strategy came Youth Demand. I have been arrested four times with Youth Demand. In November 2025, I face my biggest Crown Court trial for the Just Stop Oil gantry actions where I am expecting to receive a prison sentence of at least two years.

MOTIVATION
My primary motivation for joining the resistance is the climate holocaust that we are facing. It is now well-established that we are facing the prospect of over one billion climate refugees in the coming decades. This is only for starters – global financial collapse, global food system collapse and the deaths of hundreds of millions are in the pipeline and almost certain to happen if we continue emitting carbon at the current rate. We have known this for decades and still emissions are rising. It's obvious to me that this is the greatest evil in human history.

I have been blessed to be born in the UK to a middle-class family and have had the privilege of a good education. I live in the sixth largest economy in the world and our actions can have a material impact on stopping this horror. So, to me it is simple. I feel like I have an absolute responsibility to do everything in my power to stop this from happening; anything less is complicity. If that means going to prison for years, so be it. People have died for much less.

INSPIRATION
The person that has unquestionably inspired me most in my resistance is Roger Hallam. I am unbelievably grateful for Roger's leadership at this time. I think it is insane how few people are willing to clearly and bluntly say what my generation faces in the coming years and what we need to do about it – engage in civil resistance. Roger is one of the few leaders, if not the only one, who is saying this openly and showing intellectual leadership on what must be done.

Once I began educating myself on civil resistance, I have been greatly inspired by the Civil Rights Movement – Martin Luther King, Ella Baker and others. I have been inspired by the Suffragettes and Emmeline Pankhurst. I have also been inspired by the revolutionary movements of the recent past – Otpor!, for example, under the leadership of Srđa Popović.

IN NATURE
I honestly don't have a strong connection with any particular natural environment – living in London makes this difficult! But I definitely feel most at home in the woods – losing myself in a forest is something I love to do.

QUOTATION

One of the quotes I keep close is the following saying, seemingly often misattributed to Edmund Burke: 'The only thing necessary for the triumph of evil is for good men to do nothing.' It seems, in fact, to derive from: 'Bad men need nothing more to compass their ends, than that good men should look on and do nothing' – words spoken by utilitarian philosopher John Stuart Mill in an address in 1867. It is not the bad people of the world that create evil; it is the good people who stand by and let it happen. I keep this in my mind all the time, and I think it applies especially to the 'progressive' movements, particularly the vast majority of the 'climate movement' that has failed to lead people into resistance, and has therefore facilitated this horror.

RESOURCES

The top resource I would recommend to young people who want to understand what must be done is Roger Hallam's 2021 lecture 'Advice to Young People as They Face Annihilation'.

WHAT LIES AHEAD?

The ultimate success that we must work towards is enacting a progressive, democratic revolution in the UK. Nothing short of a revolution in our political system, which can facilitate a revolution in the economic system, will be sufficient at this point.

Of course, this is a project of the next decade or two. In the next two years, success looks like taking steps towards building the most powerful movement this country has ever seen in order to enact this revolution.

The final thing I want to say here is that there is another dimension of success which is crucial. Each one of us needs to step up to do what these times require of us, which is to resist. We must engage in actions that are proportionate to warn the country of the horror that is coming, so that we can lead people through this madness as best as we can. At the end of the day we cannot control the outcomes of our actions – we can only act with courage, honour and dignity against the evil we face. That is what true success looks like to me.

SEAN IRVING

I am a twenty-six-year-old community organiser in Co-operation Hull and a PhD candidate at the University of East Anglia. In my research I am interested in how we can transition to economies and political systems that serve everyone's needs, especially those often excluded from mainstream 'environmentalism'. I see confrontational activism as one essential – though on its own, insufficient – tool in this struggle.

Within Co-operation Hull, most of my time is on a project that is looking to develop a Community Energy Company for Hull and our weekly pay-what-you-can pop-up restaurant Waffle.

I grew up in rural Cambridgeshire in a middle-class household near a nature reserve, and was involved in the Woodcraft Folk, a social-justice oriented outdoorsy youth organisation. I went to university at eighteen to study chemistry, but came out with a degree in environmental sciences. Learning about the environmental crisis as part of my degree and seeing my sister campaign as part of the Youth Strike movement 'politicised' me. This politicising took the form of orientating my life from a standard career/family path to trying to do what's necessary to minimise the damage of our accelerating eco-social injustice and social collapse in the twenty-first century.

INVOLVEMENT

Just Stop Oil – took actions and gave recruitment talks. Most notable actions were painting climate-denying think-tank Policy Exchange orange, and disrupting an opera at Glyndebourne opera house.

Co-operation Hull – Catalyser, or three-days-a-week volunteer, contributing to the organising of People's Assemblies and the Solidarity Economy across Hull. UEA Biodiversity and Climate Action Network (BCAN) – I was involved in the post-Covid rekindling of the student-staff climate-action network at the University of East Anglia. Cambridge Climate Justice (previously known as Cambridge Zero Carbon) – student campaigning.

MOTIVATION
The deep injustice of our political-ecological systems.

INSPIRATION
I'd like to pick a figure from some great historical resistance movement, but if I'm honest, then it's my sister.

IN NATURE
Under the big, old oak tree, which is more than 600 years old, in the nature reserve which I spent so much time in near where I grew up.

QUOTATION
'Hope is not a lottery ticket you can sit on the sofa and clutch, feeling lucky. It is an axe you break down doors with in an emergency.'
Rebecca Solnit

RESOURCES
It would depend on the person, what they are interested in, and what they are already doing. One of my go-to favourites is the film Finite: The Climate of Change by Rich Felgate.

WHAT LIES AHEAD?
If I'm honest, I don't think we're going to see that much change in the global geopolitical system in the near future, unless we have a massive, unexpected shock and the Left is prepared with a co-ordinated response, both in terms of media narrative and material support for those suffering.

In my current work I would like to see a community energy sector more orientated towards autonomy, local resilience and wealth redistribution, and self-organising institutions emerging across the North of England providing for people's basic needs in an ecologically sound and resilient way.

EXTRACT FROM INTERVIEW
One thing I think we all realise is that we need to go a lot deeper in terms of our analysis. The decision taken to set up Umbrella in January 2024 was part of that, including the decision to focus on establishing Citizens' Assemblies through the Assemble initiative. It's so easy to get overwhelmed by the scale of the change that is needed, and for me that means focusing on places where we can make a real difference – which is what I am doing with Co-operation Hull. My PhD is focused on changing economics for the better, so that people can provide for their needs without the endless focus on more consumption, more growth, more waste. That's the academic end of it, but my main solution is to lean into this where I live here in Hull, to make all that stuff work in practice. How do we build social and emotional resilience, creating better spaces and communities, to substitute for so much of the pseudo-satisfactions of consumerism practice?

ACKNOWLEDGEMENTS & CONTACTS

ACKNOWLEDGEMENTS

Publishing a book in this way has been quite an experience, and I'm hugely grateful to all those who've made it possible. Particularly to Anthony Eyre at Mount House Press, for the original leap of faith, and Duncan Proudfoot, whose deep knowledge of the publishing world was invaluable; to Helen Mockridge and Carla Dobson Perez for all their help with marketing and publicity; to Mel Carrington at Just Stop Oil; to Becky Burchell and Sophie Austin for all the inspiration early on; and to Sam Johnstone, my VA, for keeping everything moving along all the way from the start!

I'm also very grateful to the Sheepdrove Trust, the Marmot Charitable Trust, the Polden Puckham Charitable Foundation and the Aurora Trust for their financial support.

CONTACTS

Defend Our Juries – https://defendourjuries.org/
Extinction Rebellion (XR) – https://extinctionrebellion.uk/
Climate Action Support Pathway (CASP) – https://mycasp.com/
Rebels in Prison Support – https://rebelsinprison.uk/
Green New Deal Rising – https://www.gndrising.org/
Fossil Free London – https://fossilfreelondon.org/
This is Rigged – https://thisisrigged.org/